The Power of Sustainable Development in Vietnam

The **ISEAS – Yusof Ishak Institute** (formerly Institute of Southeast Asian Studies) is an autonomous organization established in 1968. It is a regional centre dedicated to the study of socio-political, security, and economic trends and developments in Southeast Asia and its wider geostrategic and economic environment. The Institute's research programmes are grouped under Regional Economic Studies (RES), Regional Strategic and Political Studies (RSPS), and Regional Social and Cultural Studies (RSCS). The Institute is also home to the ASEAN Studies Centre (ASC), the Singapore APEC Study Centre, and the Temasek History Research Centre (THRC).

ISEAS Publishing, an established academic press, has issued more than 2,000 books and journals. It is the largest scholarly publisher of research about Southeast Asia from within the region. ISEAS Publishing works with many other academic and trade publishers and distributors to disseminate important research and analyses from and about Southeast Asia to the rest of the world.

The Power of Sustainable Development in Vietnam

Environmental narratives, NGOs and the state's environmental rule

JULIA L. BEHRENS

ISEAS YUSOF ISHAK INSTITUTE

First published in Singapore in 2025 by
ISEAS Publishing
30 Heng Mui Keng Terrace
Singapore 119614
E-mail: publish@iseas.edu.sg
Website: http://bookshop.iseas.edu.sg

ISEAS Library Cataloguing-in-Publication Data

Name(s): Behrens, Julia L., author.
Title: The power of sustainable development in Vietnam : environmental narratives, NGOs and the state's environmental rule.
Description: Singapore : ISEAS – Yusof Ishak Institute, 2025. | Includes bibliographical references and index
Identifiers: ISBN 978-981-5203-47-9 (soft cover) | ISBN 978-981-5203-48-6 (ebook PDF) | ISBN 978-981-5203-49-3 (epub)
Subjects: LCSH: Environmental policy—Vietnam. | Sustainable development—Law and legislation—Vietnam. | Non-governmental organizations—Vietnam. | Environmentalism—Social aspects—Vietnam.
Classification: LCC GE190 V5B42

Cover illustration by Duong Nguyen
Cover design by Lee Meng Hui
Index compiled by Sheryl Sin Bing Peng
Typesetting by International Typesetters Pte Ltd
Printed in Singapore by Markono Print Media Pte Ltd

For the Vietnamese women who cannot be named in this book

Contents

Preface

Writing a book—going from research to initial manuscript to final publication—takes time. While I was going through that process with this book, the political context changed significantly.

I set out to understand what nature and environment mean for people in Vietnam and their political system. I have worked in the broader field of ecological development and political cooperation, and I have often wondered how the cooperation partners in Asia whom I corresponded with understand the issues surrounding nature and the environment. As it turned out, these partners were less interested in international development cooperation and more in the local entanglements with different entities of the Vietnamese government. As a result, I ended up writing about sustainable development and power structures in Vietnam.

The Vietnamese environmental governance that my informants were concerned about has caused many new worries since 2021 (when I concluded my research) for Vietnamese non-state actors, international organizations and even departments within the Vietnamese government. The last three years have seen further restrictions in the opportunities to engage in political and environmental spaces. The economic decentralization that has been central to the state's governmental setup has continued to lead to decisions being made at the top state level while putting responsibility on the individual citizens. As individual clean-up campaigns and propaganda posters that encourage citizens to keep their environment clean are still going strong, prominent advocates in the environmental field have been incarcerated, among them are persons who have inspired and helped me with this research. The arrests sent shockwaves through the environmental community, with the grapevine full of speculation about why they happened. In the wider context, an understanding of the Vietnamese state's approach to governance is crucial to the bigger picture. I have not rewritten the manuscript to fit new changes since they have happened after my fieldwork. They are, nevertheless, important to bear in mind, and I have added paragraphs to point that out wherever appropriate.

The arrests and newly issued decrees were followed by anti-corruption campaigns and an increase in surveillance—for example, the staff of the restaurants where I reserved tables for dinners would always seat me at a convenient table and started to have pictures of me. But even before all that, doing research in Vietnam was challenging, especially with the lists of permits and documents required. Additionally, I stayed in Vietnam during the COVID-19 pandemic, which was a difficult period to conduct research due to the difficulty of travelling to the provinces outside of Hanoi.

I hope this book still achieves what it set out to do: generating and translating knowledge about environmental rules and sustainable development in Vietnam. This means deepening the understanding of what sustainable development means in Vietnam, why it means what it means and what conclusions we can draw from the results to better understand the functions of the state, non-governmental organizations (NGOs) and international development cooperation. The book gives an insight into not only why narratives matter, but also why and how they deviate from the practice in some places. For development practitioners, the book aims to help them better understand the context they are in and what (un)intended consequences their actions might (not) have. Across all interviews and observations, I found Tania Li's famous book *Will to Improve* continues to be relevant today. Having worked for think tanks myself, I am only cautiously optimistic that things will change for the better. Nevertheless, I am thankful to the people you are going to meet in the following pages.

Acknowledgements

This book is the product of my long PhD journey. The journey has been especially long because it took place during the COVID-19 pandemic and while I was moving between countries and having many self-doubts. To be able to complete my research and publish this book, I owe it to several wonderful, inspiring and supportive colleagues, friends and family members.

In 2015, I started on my PhD journey with a different research topic from what it is now and under a different supervisor. I am very grateful to Professor Dr Michael Mann for encouraging me to take my first step towards a PhD. I am not sure if I would have embarked on this endeavour without his support and advice. Along the way, I changed my research topic, and I am most grateful to Professor Dr Vincent Houben for accepting me as one of his PhD students and giving his precious time and expertise. He has supported me not only on my academic journey, but also on my professional journey with his countless approvals, comments and letters of recommendation. Professor Dr Christoph Antweiler has provided meaningful comments to my initial thesis, and these helped me to strengthen my argument for publication. During my time in Vietnam, I would not have been able to settle down and navigate academics without Professor Dr Pham Quang Minh's help. Despite his busy schedule, he always makes time for his students. There are also several other academics and scholars who supported me with their advice, contacts and encouragement when I was not sure if I should continue my PhD journey. I would like to thank Professor Dr May Tan-Mullins, Professor Dr Ben Tran, Professor Dr Christina Schwenkel and Professor Dr Erik Harms. Professor Dr Harms (and Sarah Affenzeller) enabled me to spend precious time at Yale University, despite all the other global matters (a pandemic!) going on. This period was crucial for me to organize my thoughts and put them together in my analysis. I would also like to thank everyone at the now defunct Nordic Institute of Asian Studies for having me during an intense period of writing and the Asia Research Institute in Singapore for a helpful graduate workshop. Professor Dr Minh Nguyen has enabled me to dedicate my time to writing through a fellowship. She, together with Professor Dr Antje Missbach and their teams at Bielefeld University, has challenged my arguments and expanded the contribution that I hope to make.

I am very privileged and lucky to have received the support from some of the most incredible women whom I met during my fieldwork. I dedicate this book to them. Their wisdom, trust and honesty carried me through my research.

I would like to express a huge thank you to everyone who dedicated their time to reading parts or all of this research and providing me with their comments and feedback. Marie Carnein, Julia Holz, Isabelle Windhorst, Mandy Ma—I hope you know how much your contribution mattered for my dissertation. Vito Dabisch spent hours with me online, writing our dissertations together and keeping each other motivated during the lockdowns. The two anonymous reviewers played a crucial role in guiding me to structure my thoughts. Ng Kok Kiong and everyone at ISEAS Publishing have been a real delight to work with, and their professionalism have made this book readable.

Finally, I would like to thank my husband, Philip Degenhardt; my parents, Katrin and Peter Behrens; my partner in crime, Dieu Linh Dao; and my number one cheerleaders, Lis, Mai and Phung, for their mental and material support throughout the process. All remaining errors are my own.

1

INTRODUCTION

In 2011, I was riding my motorbike and stopped at a traffic junction in Hanoi. I was then studying Vietnamese as an undergraduate student in Vietnam. As the traffic light was red and the drivers were patiently waiting for the traffic flow from the intersecting street to die down, a few students with raised posters and uniformly dressed in "350.org" t-shirts, bravely crossed the street in front of the motorbikes. The words on the posters read "*Đèn đỏ tắt máy*" [Red light, turn off your engine]. Having been involved with the environmental movement in Germany since 2005, I was intrigued, to say the least.

This was the first time I had personally witnessed public environmental action in Vietnam. The incident happened at a time when the 350.org movement gained momentum and spread worldwide, transforming many young people into activists. At the time, I was not aware that this global movement had spread to Vietnam. I was fascinated at what this movement meant in different places and how actions were adjusted to different contexts.

I continued to stay in touch with environmental student groups throughout my undergraduate years in Hanoi. Even after I have graduated from university, I remained interested in environmental movements in Southeast Asia. After graduation, I got a role as a coordinator between the head office of a German political foundation in Berlin and country offices in Southeast Asia. While

I was working with the people and organizations in Southeast Asia, a few questions came to my mind: Are we actually dealing with the same problems from all perspectives? Do the solutions we are jointly working towards make sense? How much must the project partners in Southeast Asia perform in order to get funding?

What remains true across borders and boundaries is that in times of environmental crises, immediate and effective actions are necessary to save lives and livelihoods. Environmental projects are needed not only to mitigate crises, but also to adapt to unavoidable consequences. Yet, in a global context where resources are unevenly distributed and the needed actions rely on projects and funding from another place, a few questions remained: Are the projects that exist necessary and effective? Or are power relations and misunderstandings of sociocultural translations hindering truly meaningful actions? This book cannot provide answers to all these questions. Rather, it begins by examining environmental action at a specific place and what narratives frame this action and the reasons behind it. This research maps environmental actors and their actions in Vietnam by focusing on the relations between the state and non-governmental organizations (NGOs). It traces how the narrative—Sustainable Development—is used to build environmental rule, why the Sustainable Development paradigm is used by different actors to navigate state-society relations and illustrates how the development practice puts cracks in the ecological modernization programme of the narrative.

Environmental actors in power are not homogenous and are not located at only one place. They are positioned across the trans-scale processes of environmental policies and actions. Central to this is the Vietnamese authoritarian state. International organizations (e.g., head offices of international NGOs, donors and cooperation partners of Vietnamese NGOs) also exert power in discourse setting and development agendas. NGOs, the cornerstone of this research, make trade-offs between their own definition of causes of ecological crises and the possible solutions. They navigate the red lines set by the Vietnamese state and funding guidelines, bringing together external expectations and their own theories of changes. They adjust their positionality and roles to what is possible and necessary. They adopt certain languages and then act accordingly. How the frictions between practice, narrative and perceptions in the end look like is not equal between organizations. The diversity in environmental actions of NGOs in Vietnam informs us about the state's environmental rule.

Decentralization and the norms and values created are an interlinkage of capitalist market-based economy that forms the basis of the state's ecological modernization programme. This programme is framed in Sustainable Development and socialist morality that encourage citizens and institutions like NGOs to contribute to the development of their nation. In this process, the Sustainable Development narrative becomes universal and is co-created by a diversity of actors. In this discourse making, the Vietnamese government takes on a central role, and it uses the narrative strategically for its environmental rule. But the decentralization efforts make space for actors to co-shape the narrative through practice.

While Chapter 2 sets the scene for the book and explains the Vietnamese context, Chapter 3 discusses whether NGOs and civil society organizations exist in Vietnam. There is a range of research on non-governmental actors in the Vietnamese context, for example Hannah (2007), Wells-Dang (2011), Salemink (2011), Wischermann (2003, 2010, 2018) and Bui (2013). These research debates on the existence of "non-governmental" in Vietnam and the definition of civil society and non-governmental organizations. Kerkvliet (2019) has coined the term "responsive-repressive" to describe the state's interaction with other actors, summarizing how the socialist one-party state navigates dissent and conflict within the country. The book describes the interaction between NGOs and the Vietnamese state in environmental governance by focusing on three perspectives: self-identification by the NGO staff members, the official definition by the state and the networks that the NGOs positioned themselves in. I argue that approaching NGOs through these three perspectives helps in understanding the state-society relations.

Chapter 4 takes us beyond the Vietnamese state by outlining cross-scale processes that inform environmental rule. It portrays international flows that have shaped environmental narratives throughout history and in contemporary Vietnam. It shows how NGOs are embedded not only in national authoritarian power relations, but also in international ones from the colonial times. Nevertheless, NGOs have spaces to navigate in both narrative and practice. In addition, power relations within NGOs persist, too, and open up the process analysis towards local entanglements. For example, the urban-based, middle class shapes organizations, perpetuates power relations regarding ethnic minorities and marginalizes socioeconomic groups. Chapter 4, therefore, reminds us to look at processes beyond scales for meaningful research.

Chapter 5 analyses in detail the Sustainable Development narrative and compares it with the actual practice of environmental action. The difference between narrative and action shows that we need to look at both discourse and practice to get a realistic understanding of development processes. The research, therefore, reunites the opposing ends of the development discourse as exemplified by Ferguson's (2005) and Mosse's (2005) contributions on understanding development. Each of the chapters is accompanied by ethnographic stories that illustrate the arguments in the chapter through a case study.

Data and Limitations

The data in this book is based on the discourse analysis of Vietnam policy papers, laws and strategies, NGO publications, twenty-nine qualitative interviews with NGO representatives and key experts, and participant observation and focus group discussions in three case studies of NGO projects. All data were obtained between summer 2019 and spring 2021.

This time frame has three limitations for the research. Firstly, the 13th Party Congress of the Communist Party of Vietnam (CPV) was held in spring 2021. This event curbed the opportunity for research permits, especially for interviews conducted outside of Hanoi. In the run-up to the Party Congress every five years, the political situation in Vietnam comes to a standstill because new leadership is elected and all state-affiliated officials wanted is to secure their positions. NGOs postpone projects during this period, and research permits are not granted. Secondly, I undertook a major part of this research during the COVID-19 pandemic. While I have been lucky enough to have been in Vietnam prior to the shutdown of the country, the lockdowns nevertheless made it difficult to conduct field research. Trips occurred on short notice, were often rescheduled, postponed or cancelled. Thirdly, research for this book took place before the crackdown on NGOs in 2022. The text has been updated where appropriate, but the data were obtained before the change of political climate, so I have left out some information regarding my informants for ethics and security reasons.

Another bias to the research is my own positionality. I come from a background of NGO work and being a white Western woman comes with its privileges as well as its limitations. Many NGO representatives were willing to talk to me while some community members were hesitant to tell me about the challenges they faced. Being a woman who is conversant in Vietnamese language gave me advantages: the three female NGO leaders in the case studies were willing to support my research because of my language skills and because I am a fellow woman. Interestingly, my East German family background also formed informal relations with the interviewees through a common perception of socialist experience. Another crucial point is that my husband works in the NGO sector in Vietnam, and this gave me access to interviewees.

I did not interview all organizations in the environmental field; in 2017, there were 180 Vietnamese organizations whose work centred around environmental topics and were registered under the Vietnam Union of Science and Technology Associations (Ortmann 2021). I excluded those based in Ho Chi Minh City in my analysis. In Hanoi, I included various types of organizations in my sample: both large-scale and small-scale (in terms of staff strength and funding obtained); international NGOs (INGOs) and Vietnamese NGOs (VNGOs); those with different work approaches (e.g., policy advocacy, capacity building and research) and different ways of incorporating environmental concerns (this will be elaborated more in Chapter 2).

Developing a methodology to track changes in narratives throughout history and attribute them to certain actors is challenging. This methodology would have given my analysis a deeper historical perspective. However, written sources are often conceived by the educated class. Sources about environmental perception by the wider population are found in folktales and folk songs, but because they are orally transmitted and recorded in writing only recently, we can date neither the stories and their metaphors to a certain period, nor their changes. Also, it is unclear how the different belief systems in Vietnam have informed the mindsets and practices of the people. The various religions and beliefs make it difficult to date them to a certain period. Therefore, I add brief historical contexts to actors where appropriate and where existing research allowed me to do so. Despite the difficulty of establishing a sound analysis, contextualizing the history of Sustainable Development helps to establish the narrative in Vietnam.

Over ten years after my first encounter of environmental action in Vietnam, this book tries to answer the questions that I had in my mind back then. A lot of practitioners in Vietnam have sought my advice and are interested in my research results because they feel a disconnection between their projects and the reality in Vietnam. This book, therefore, is of practical significance for actors who are involved in environmental work in Vietnam. It contributes to academic literature on the environment and Vietnam from an Area Studies perspective. It examines the Vietnamese context and uses various Vietnamese voices as its point of departure for research in a system of oppression. I hope that my research encourages meaningful and effective environmental involvement in Vietnam (see also Li 2007).

Environmental Rule in Late Socialism

Environmental governance in Vietnam has been a topic in several research projects, covering how the state institutionalized environmental responsibilities in its system (Benedikter 2014; Ortmann 2017), how cooperation among different actors functions within the country (O'Rourke 2004; Zink 2013; McElwee 2016) and regionally (Wong 2012; Holzhacker and Agussalim 2019). These are important resources for my research in understanding the Vietnamese state.

Environmental authoritarianism is a concept connected to the decline of socialism in the 1980s and has been applied in different contexts (Ophuls 1973; Doyle and Simpson 2007; Beeson 2018; Arantes 2023). Beeson (2018) and Arantes (2023) applies the concept to understand sustainability and its top-down application in China. In addition to Beeson's work on political actors, Arantes describes how grassroot initiatives have become part of building a "green consensus" as a norm of environmental action in China. Many of the processes she illustrates for the Chinese context can also be found in Vietnam. Yet, I chose not to centre the analytical framework of this book around environmental authoritarianism but situate it in the concept of "environmental rule" (McElwee 2016).

In *Forests Are Gold*, McElwee (2016, p. 5) contends that the concept of "environmental rule" is useful for understanding environmental politics in Vietnam because "the concept offers a clearer explanation for the interventions directed at nature, which have not been confined to linear patterns of capitalism, socialism or neoliberalism, as others have asserted". She focuses on "unexpected relational interactions" that are constantly changing due to their context and how they transform environmental action over time. The concept recognizes that environmental rule does not come from one actor or one party alone, but from a variety of actors.

For her methodological framework, McElwee (2016, p. 14) breaks down environmental rule into these categories: problematization, knowledge-making, directing conduct and subject making. The point of departure for my research is problematization. Although all four fields matter in my analysis, my approach and focus differ from McElwee's. While she focuses on forest policies, I set out to understand the environmental narratives that frame policies in general and their role in governance so that I can contrast the discourses with the practices of development. While the linear patterns of political and economic systems are not sufficient for understanding environmental action, they nevertheless shape realities when investigating the environmental narratives closely. Therefore, the systemic frameworks of capitalist market economy, socialism as well as other cultural factors are important to my analysis. What remains similar in both our work is the emphasis on the diversity of actors that are involved in the processes and that are part of the social and cultural contexts that define Sustainable Development in Vietnam. Looking beyond the nation-state and the history of current policies is necessary for a complete understanding of environmental rule.

This book is set in the political context of the Vietnamese one-party state, led by the CPV. I chose the context of late socialism and not post-socialism for this research. Wilcox et al. (2021) and Leshkowich (2008) have, among others, described that late socialism better describes the realities in Vietnam, China and Laos. Post-socialism is used to describe the context of former socialist states (e.g., Eastern Europe, Russia, Central Asia) that continue to be impacted by their socialist past. It has also been applied to Vietnam, China and Laos to underline the realities of capitalist market-based economies and the major changes the countries have undergone over the last decades. But putting these very different sociopolitical realities into one category risks overlooking fundamental principles in power exertion. The Vietnamese state has continued to operate officially with socialist morals, propaganda, goals and policymaking. The governance that comes with this assumption of socialism is central to the analysis of this book. Therefore, I still recognize the influence of capitalism in today's nation-state and find the term "late socialism" the best fit for the reality in Vietnam.

Arriving at Environmental Narratives

The journey from Berlin to Hanoi and eventually to the location of the three case studies in Central Vietnam and the Mekong Delta has unravelled my own positionality and assumptions and made me rethink the conceptual framework that I used in this research. I was fortunate to have interviewees who did not refrain from asking me questions and thereby informing the analysis; for example, "Why do you focus so much on international actors? We care much more about what the government tells us"; "Why do you want to focus on North Vietnam? You need to come to the Mekong Delta to understand what environment in Vietnam is about". As I analysed my data and applied grounded theory to include my informants' views, it became clear that the contradictions in my conceptual framework seemed to be connected to one big goal: creating a liveable environment for all without getting banned by the Vietnamese government and without losing funding.

Consequently, I use Tsing (2005)'s theory of universality and frictions to explain environmental action by NGOs, environmental narratives and power structures in Vietnam, while I define actors in the research based on Latour (2007)'s actor-network theory. To understand the meaning of environment and nature for NGOs and how these meanings are shaped by NGOs' entanglements in power structures across scales, I look at environmental narratives following Tsing (2005)'s approach of analysing universalities and the role of different actors. I then contrast these narratives with the practices thereof in three case studies. Looking at only the discourses would be misleading because the mere use of words and concepts does not equate to the actions behind the terms. For example, an organization may use certain terms in their documents to fulfil donor and government requirements and to explain their work to outsiders. However, the actual practice in projects may differ from the discourses or they may be implemented in different ways.

I refer to environmental narratives using the concepts by Hajer (1995) and Forsyth and Walker (2008). Hajer (1995) equates narratives to storylines and discourses that are the results of negotiations between different actors and their framing of a problem. Instead of using clear and uncontested definitions of words, he understands environmental terms as concepts that define the reasons and possible solutions to a problem. Additionally, looking at the narrative and deconstructing which actors are involved in its making tell us about the political agenda behind it. All contexts of time, space and sociocultural relations between the actors informing the narrative make up a discourse setting.

Forsyth and Walker (2008) refer to this understanding of Hajer's and extend it to their analysis of environmental change in Thailand. The authors seek to overcome the dichotomy between local and scientific knowledge and between local and state knowledge, and instead show how environmental narratives are made and remade by different actors that influence each other in their knowledge production. Therefore, they seek to break down hierarchies and power structures involved in making the narrative.

Knowing what constitutes environmental narratives helps us to understand their meaning and function in the research context. Structuring environmental problems in this way also means assigning accountability to certain actors (Hajer 1995), leading to a simple understanding of environmental change (Forsyth and Walker 2008). Hence, narratives fulfil an important political function by valuing certain storylines and experiences more than others and drawing practical consequences from environmental accountability, thus putting actors in fixed roles and positions; "in other words, environmental knowledge and social order are coproduced" (Forsyth and Walker 2008, p. 18). For this research, I am interested in studying this coproduction and not in the soundness of environmental narratives and the descriptions of environmental crises. Instead, I seek to map out the different understandings of problems, solutions and responsibilities to make sense of NGOs' work and the environmental rule that is informed by multiple cross-boundary processes and negotiations between actors.

Due to Sustainable Development's omnipresence and the various actors involved, this research focuses on its deconstruction and explains how it became universal in Vietnam. Some alternative discourses are also discussed. All narratives are analysed according to the problems and solutions as well as the responsibilities and blames that they focus on. This brings up several frictions which I will point out. Additionally, I give this analysis a historic perspective and retrace the agendas that the actors used to interpret the narratives.

Sustainable Development

Sustainable Development—the central concept of this book—is omnipresent and important for environmental governance in Vietnam. It does not originate from Vietnam, but it has made its way into the country from the international policy sphere. The Sustainable Development concept is dated to 1987 when the World Commission on Environment and Development, also known as the Brundtland Commission, defined the term in its report as a global response to the looming environmental and humanity crises. It defined Sustainable Development as "development that meets the needs of the present without compromising the ability of future generations to meet their own needs" (United Nations General Assembly, 1987, p. 43). Before that, Sustainable Development was occasionally used as a term but without a clear definition; it was contextualized from case to case before 1987. The concept is embedded in the development discourse.

Escobar (1995, pp. 194–95) establishes a link between the traditional development paradigm and Sustainable Development by drawing parallels: "again the global is defined according to a perception by those who rule it. Liberal ecosystems professionals see ecological problems as the result of complex processes that transcend the cultural and local context'" and "the Western scientist continues to speak for the Earth. God forbid that a Peruvian peasant, an African nomad, or a rubber tapper of the Amazons should have something to say in this regard". This analysis is too short sighted; as we shall see in the practice of Sustainable Development in Vietnam, peasants, workers, NGO staff and others have an agency to fill the narrative with practical meaning and work towards development on both structural and individual levels.

As Bernstein (2005, p. 659) argues, the Brundtland Report provides the first legitimation of the environmental protection-economic growth nexus with an emphasis on technocratic management. According to Escobar (1995, p. 192), this nexus was a linear next step from the development paradigm and combines the eradication of poverty with protection of the environment and Western hegemony through its concept of the environment and economy. Both poverty and environmental destruction were not perceived as inherent in the economic system, but in the poor management of the political-economic system, which could be fixed with technocratic reforms. With the right economic policies, people could be lifted out of poverty and adopt an environmentally friendly behaviour. Despite the Brundtland Report establishing a path dependency within this environment-economic nexus, it still left space for interpretation as it did not suggest methods on how to tackle environmental crises in detail (Tulbure 2011, p. 126). This space was intentionally created to frame Sustainable Development as a matter of consensus, despite the pluralistic concerns that vary globally (Grunwald 2011, pp. 19–20). The Vietnamese government used this space for its authoritarian approach to environmental rule and reinforcement of local power structures, while following the environment-economic nexus. The nexus shows that Sustainable Development is not completely content-free (Grunwald 2011), but it has developed a basic common ground for actors to agree upon when referring to Sustainable Development.

In academic literature, several authors have defined the Earth Summit in 1992 as another key event in strengthening Sustainable Development on the global level. Bernstein (2005, p. 659) claims that it was this summit that the Sustainable Development discourse took on a route of "economic liberalism" as it "institutionalized the view that trade and financial liberalization, and corporate freedom, are consistent with, even necessary for international environmental protection and sustained economic growth". Critics see this summit as a failure because it did not create a joint strategy and was too focused on enhancing economic growth; the singular focus on economic growth overshadowed the aim to stop climate change (Degenhardt 2016, 5f). Principle 12 of the Rio Declaration that was produced at the Earth Summit provides the crucial point of this critique. It says:

> States should cooperate to promote a supportive and open international economic system that would lead to economic growth and sustainable development in all countries, to better address the problems of environmental degradation. Trade policy measures for environmental purposes should not constitute a means of arbitrary or unjustifiable discrimination or a disguised restriction on international trade [...] (United Nations General Assembly 1992).

This political commitment is one example for how Sustainable Development has become an ecological modernization programme. It is itself proof of the emergence of an international regime. Additionally, we can see the rise of importance of non-state actors, especially private businesses. Decentralization of policies relies heavily on the attribution of power to market mechanism and seeing them as crucial instruments for environmental policies. In this process, privatization has gained prominence, and it uses science and technology to categorize ecological crises.

Ten years after the Rio Summit in 1992, the World Summit on Sustainable Development took place in Johannesburg and continued the environmental narrative of solving ecological crises using an ecological modernist agenda. This includes the promotion of public-private partnerships and privatization, making private businesses a central actor in environmental action (Bernstein 2005, p. 660). Even when the UN World Summit confirmed its view on the environmental, economic, social and cultural pillars of Sustainable Development, the hegemony of capitalist ideas of modernity remained and influenced all pillars (To 2011). The Sustainable Development concept was developed through different blueprints, and it was first originated in connection to the Green Growth concept in 2005 at the Fifth Ministerial Conference on Environment and Development (MCED) in Seoul. The Green Growth concept was an idea derived from an Asian initiative, and it expanded to international and multilateral organizations and to national governments. In 2012, Green Economy became a key term at the Rio+20 United Nations

Summit (Jacobs 2013, pp. 197–98). The term refers to Green Growth debates that have been popularized since about 2008 by international economic and development institutions, such as the World Bank (Jacobs 2013), which play an important role in the Vietnamese government's understanding of an authoritarian socialist state with a market economy. The ambiguity of Sustainable Development continues to leave space for different forms of environmental rule and policymaking.

As another milestone in the global making of Sustainable Development, the Sustainable Development Goals (SDGs) were proclaimed as successor to the Millennium Development Goals (MDGs) by the UN in 2015. The SDGs function as a framework for policies across the globe. Holzhacker and Agussalim (2019) argue that the SDGs have the advantage of being more qualitative and less quantitative than the MDGs, therefore making space for more local solutions. They are a "shared language for engaging contested futures" (Swilling 2019, p. 3). Economic growth and capital accumulation remain explicit in Goal 8: "Promote sustained, inclusive and sustainable economic growth, full and productive employment and decent work for all" (UN website, accessed 20 November 2021). The language of the SDGs is not all-inclusive, but ecological modernist. In the case of climate change, this means, for example, that the narrative of climate change is based on a modernist natural science understanding and framed in numbers (e.g., carbon dioxide emissions, carbon footprints), and problems can be solved using innovative techno-fixes and market regulations (Lindegaard 2020, p. 159). Consequently, the dependency on partly regulated market economy is still in force, and this narrows the field of possible environmental action and with it, actors who partake in this environmental narrative. The economic system and its strong actors exclude alternative visions and those actors who try to establish a newly regulated economy outside the growth narrative.

We shall see later in this book that the SDGs and the Paris Agreement play a major role when connecting activities in Vietnam to the international scale. They serve as a broad context and reference for specific goals to which organizations contribute their work. While the importance of the Paris Agreement is limited to organizations that work on climate change, the SDGs tend to be universal so that all organizations find some points of connection.

The ambiguities of the Sustainable Development narrative in general, and the SDGs in particular, make space for different interpretations. As Tsing (2005) notes, this global scale discourse has become universal by its entanglement and transformation through localities and their influence on the discourse as we shall see in Vietnam. Sustainable Development invites different actors to fill it with practices and meanings, making it universal. Vietnam has picked up all the abovementioned key terms and discourses and translated them into national policy with some local adjustments. At the same time, the Vietnamese government has redefined Sustainable Development for their own means and has only passively accepted international paradigms. The restricted framing of Sustainable Development is welcomed by the state, which uses the global discourse to further power structures. The CPV uses it to maintain their legitimization as a problem solver and to characterize alternative narratives as a threat to the environmental rule.

Environment and Nature—Environment or Nature—Environment vs. Nature?

Research on the sociocultural understanding of environment-human relations in Vietnam is rarely undertaken. Especially in the field of political science, scholarly works assume, for example, that the Sustainable Development is used to analyse frameworks of environmental policy. Analyses on the narratives used in Vietnam on what the environment is and where the understandings come from, are constituted on ethnic minorities (Lundberg 2004; Salemink 2011), but they do not explicitly question understandings of the Kinh ethnic majority. My research is concerned with environmental narratives and power structures in NGO work in Vietnam, and I therefore add to the existing literature with contemporary perceptions of ecological crises and the definition of problems and solutions thereof.

In seeking to understand Sustainable Development, the readers of this book need to know what I meant by terms like "environment" or "nature", which have sociopolitical meanings. The term "nature", caught in a net of dichotomies, is a highly contested one. The opposite of the term "nature" has been described as "culture". Still, in today's understanding of Western science as well as Western literature and arts, nature is described as something "natural" that is not touched by "humans"; it is a state that has existed before culture (Latour 1993). For example, a patch of grassland can grow naturally or be cultivated into a garden. In colonial history, colonized communities have been described as closer to nature, reinforcing racist stereotypes but also self-perception of white colonizers. Tsing (2005) describes, for example, that scholars in service of colonial powers conducted botany such that the gathering of non-European plants also defined European consciousness about themselves and their modern, strategic thinking. The nature-culture dichotomy does not only influence thoughts in the Global North. Throughout the chapters in this book, it will become clearer that actors in Vietnam share a similar understanding of contrasting nature and culture against each other. Nature continues to be defined not only as "non-human" in a passive way, but also as an actor. Phrases like "nature hits back", "nature is getting revenge" (quotes from my interviews) and more implicit descriptions of how nature makes people calm and healthy portray nature not as a passive territory, but as an actant.

Nature and environment: What is the actual difference between these two? All interviewees, no matter whether they are Vietnamese or not, working for INGOs or VNGOs, agreed that humans are the decisive factor in differentiating between the two terms. Environment is everything surrounding a person or a thing; it is defined from the perspective of a subject. It includes water, air, trees, natural resources, humans and everything man-made. In contrast, nature is everything that "has existed from the beginning" without humans. This dichotomy in thinking seems to be very close and relatable to traditional European environmental philosophy but should not be traced back exclusively to colonial flows. Instead, roots in emic historic concepts as laid out in Chapter 2 need to be considered too.

Latour (1993) combines the nature-culture dichotomy into one term—"nature-culture". He characterizes "nature-culture" as a construct of "humans, divinities and nonhumans". In his later work in 2017, Latour was critical towards this conceptualization. Although the term describes the modern living concept and its shortcomings, the divide between humans, divinities and nonhumans seems problematic when we move outside of the Western sphere. In the Vietnamese context, the difference between these three can become quite blurry. Hồ Chí Minh, for example, was a human being but is now treated as a divine in some parts of the country (Lauser 2008). The category of "nonhumans" is questionable as trees, stones, etc. can be seen as reincarnation or symbols of human beings. Nature needs to be seen with an understanding of the actors in it. This means that nature is separate from humans and has its own agency.

The other ambiguous term—the environment—also needs to be viewed critically. Escobar (1995) describes the environment as a transformation of nature to rid it of agency and to fit it into an economic understanding. This is, however, not true for all environmental narratives and for all actors, and it focuses too much on a particular understanding, leaving out resistance, contestations and alternative understandings. I agree that the environment lacks agency, but not because it is rendered passive. Instead, I understand the environment as a space in which humans, nature and other actors interact. This definition underlines the social and political elements that are essential to the concept of environmental narratives and the analysis of what Sustainable Development means for different actors in Vietnam and why it or alternative narratives are used.

What is Vietnamese?

The Vietnamese nation-state is multiethnic and draws from various cultures. Ethnic minorities in Vietnam do not necessarily perceive themselves as "Vietnamese". Discrimination against ethnic minorities has been and continues to be widespread. The ethnic majority, the Kinh or Việt, are the focus of this study, and to underline the cultural strains that have predominantly existed among them, I will speak of Kinh or Việt culture, instead of Vietnamese. The power relations between Việt and ethnic minorities are highly relevant for NGOs and their projects. As Salemink (2011, p. 48) phrases it, "historiography and ethnography of Vietnam also require a view from the mountains in order to redress the nationalist and developmental notions about backwardness, remoteness and isolation produced by the modern state and eagerly supported by NGOs and other development donors." Ethnic minorities use environmental narratives in the othering process during environmental actions (this will be discussed in detail in Chapters 3 and 4). In the next chapter, I shall introduce the actors of this book and the context of Vietnamese governance.

2

Environmental Rule: Meeting Actors in State-Society Relations

There is no way around a thorough contextualization to avoid some common mistakes: homogenization of complex sociopolitical relations, historicalness and overly dichotomization. This chapter seeks to counter these three analytical oversimplifications through a detailed discussion of the actors central to this research, their relationship to each other and their overall embeddedness in the environmental governance present in the Vietnamese context. It then demonstrates how historic events have impacted the understanding of human-nature relations and how environmental narratives have evolved. Studying the past is crucial to understanding why particular narratives that seemed opposing, are depicted as complementary by state actors. Lastly, the chapter not only takes a closer look at the state-society relations through scrutinizing relevant actors in environmental governance, but also recentres the relations between humans and nature by discussing landscape as a consequential creator of sociopolitical contexts.

Relevant Institutions for Environmental Policymaking in the Vietnamese State

The socialist republic is under the leadership of the Communist Party of Vietnam (CPV), which is the only party allowed. The authoritarian rule in this one-party state is structured horizontally from the national level downwards. This means that the CPV has a system in which the inner-party organizational structures run parallel to the actual governing structure throughout the whole country. The elected people's committee works alongside the level's party secretariat. The lowest governing level is the village (*thôn*) and hamlet (*xóm, áp*). Next is the commune level (*xã*) or, in urban context, small towns (*thị trấn*) or wards inside a bigger city (*phường*). This is followed by the district level (*huyện* for rural areas; *quận* for urban areas) and the provincial town (*thị xã*); provincial level (*tỉnh*) and cities (*thành phố*); and finally, the national level (*trung ương*). Authorities are responsible for what is happening in their areas, and they report to the next higher level. The direction of reporting is supposed to create a bottom-up approach, but in practice, directives and decrees from the upper level are handed down and applied on the lower levels (Ortmann 2020). Sometimes local authorities (i.e., up to the provincial level) adjust directives handed down from above to their own context. But in many cases, cadres lack either capacity, knowledge or both to do so and simply copy the text handed down from the national level and insert their area's name. In addition to party and government structures, there are mass organizations, such as the Vietnamese Fatherland Front, Vietnam Farmers' Union and Vietnam Women's Union, which are supposed to represent people with different interests and include them in the overall CPV structure. The Party is, by law, the only legitimate representation of the people.

Both government and party positions are filled through a process of selection and election (Kerkvliet 2004). The Party shortlists candidates for a position, and the people are to vote from this selected list. Although the people get to elect their leaders, the choice is still limited to those candidates approved by the Vietnamese Fatherland Front. Persons whose loyalty to the CPV is unclear are not selected. A famous case was in 2016 when pop singer-turned-dissident Mai Khôi declared that she wanted to run for the National Assembly. However, she was not shortlisted. In response, she started campaigning for freedom in Vietnam. Eventually, she lived in exile after receiving various threats from the CPV.

At the highest level, the Central Committee, the Politburo and the Party Secretariat are at the institutional centre of the Party's power. The general secretary is the most powerful person in Vietnam. Beside him, there are three other persons—the president, the prime minister and the chairperson of the National Assembly—that are considered the pillars of state power. Until recently, they were selected and elected every five years, shortly after the election of the National Assembly; these appointments are now confirmed during the Party Congress that takes place before the elections. Every five years the Party members come together from local to the national levels to elect new leadership and discuss their future political direction; concretely, policies and law changes for the next five years and a vision for the next ten years are introduced. The 13th Party Congress took place during this research in spring 2021. This event tends to be closely observed because it reveals the direction for the country in the coming years and demonstrates the ongoing power struggle within the party, which is by no means a homogenous block (Hiep 2020). The 13th Party Congress manifested the power of General Secretary Nguyễn Phú Trọng, but also left power struggles within the party unresolved. Not even two years after the Congress, the Socialist Republic of Vietnam has a new president since the previous one had to resign due to corruption scandals. Rumours about the unstable political future of the current prime minister keep circulating around Hanoi.

During the 1950s, along with land collectivization, the official state architecture gradually had a grip on the villages. During that time, the state recognized the commune level (xã) as the basis of the state, and the success of the CPV is founded on a party network that can reach out to the local villages, form the connection between governing levels and thereby link the local level to the national level. This system allows for villages and other actors to adjust central practices to their localities, as long as they do not endanger the legitimacy of the party or the overall network. Kerkvliet (2019) describes this system as "responsive-repressive".

Within this system, the government has set up several institutions as well as policies to tackle environmental issues. This includes the establishment of the Ministry of Natural Resources and Environment, the National Council on Sustainable Development, the Steering Council on Sustainable Development, and offices of Sustainable Development at the Ministry of Planning and Investment. Due to Vietnam's authoritarian context, it is necessary to understand the state's environmental institutions and their roles. These institutions do not just control and regulate environmental actors (the repressive part in Kerkvliet's definition), but they also cooperate and comply with them (Ortmann 2017). The environmental governance field, therefore, is "evolving" (Ortmann 2020, pp. 196–97) by including more actors while maintaining a top-down structure. Beyond environmental governance, policymaking is also closely connected to social governance. Implementing environmental policies with a social agenda is what McElwee (2016) calls "the environmental rule", and this is similar to Forsyth and Walker's (2008) analysis on the co-dependency of social and environmental regimes.

The central government actors include offices, agencies and centres that have been established to tackle environmental problems. Weller (2006, p. 138) makes an analytical observation that also applies to the Vietnamese context:

> Their work suggests that taking the "state" as a whole is a misleading reification. Even more, these complexities require us to move beyond the simple hierarchy of above and below, and even beyond an image of the state as an orderly hierarchy where power ramifies down from the top to lower branches, like descent flowing down a genealogical chart. The anthropological concept of heterarchy, which recognizes that hierarchy is complicated and compromised by units whose relative power may vary, may provide a better metaphor. Heterarchy does not result in the unified structure of an organizational chart, but rather in a conglomerate of multiple, competing orders of power and authority.

The struggle for power in heterarchy starts at the ministerial level. In 1993, the Ministry of Science, Technology and Environment was established. Around the same time, the environmental law was introduced, through which Vietnam rolled out its environmental governance strategy. However, the institutional structure has changed since then. What was previously one ministry is now split into two: the Ministry of Natural Resources and Environment (MONRE) and the Ministry of Science and Technology (MOST).

The new MONRE focuses its work on environmental legislation and execution. It is responsible for climate change strategies and policies and has established the Standing Office for Climate Change under its liability. However, its influence within the government is limited. In the hierarchy of ministries, the MONRE is below the Ministry of Planning and Investment (MPI). The MOST also plays a role in environmental policymaking due to its focus on using science and technology in solving environmental problems. But the MPI, the Ministry of Industry and Trade (MOIT) and the local governments of some provinces and cities claimed responsibility in drafting the National Target Program to Respond to Climate Change (NTP-RCC). The competition and hierarchy among the ministries and offices complicate decision-making, especially as the number of bureaucrats has risen constantly over the past decades (Painter 2006; Benedikter 2016). MONRE's main task was to coordinate all actors in the NTP-RCC framework. However, this framework turned out to be just a network-coordinating body, without a clear action plan for climate change adaptation and mitigation (Fortier 2010, p. 234). Environmental policies and strategies tend to lay out the long-term goals without the step-by-step implementation plan.

There are several key government papers that set out the state's approach to domestic environmental governance, in line with international regimes, protocols and agreements that Vietnam has ratified and/or signed. These papers formed the basis for the discourse analysis in Chapter 3. Among the first such government papers is that on Project VIE (1989), which was officially approved and adopted in 1992. Project VIE contains the "Vietnam National Plan for Environment and Sustainable Development 1991–2000", which can be seen as the starting point of Vietnam's modern environmental policymaking. It is the first document on environmental policy approved after Vietnam's *Đổi Mới*. It formed the basis of what then would become the setting up of a legal environmental framework (Nguyen 2012, p. 23).

Before *Đổi Mới* in the 1980s, Vietnamese environmental policy was characterized by war and the building of a socialist republic, whose goal was to lift its people out of poverty through industrialization and modernization. In 1972, for example, at a conference in Hanoi, the problem of deforestation and destruction of nature due to warfare was discussed. With the aim of eradicating hunger and poverty, Vietnam's environmental concerns were steadily framed into economic concerns.

The state's official narrative in economic politics shifted from a socialist ideology towards "a market economy with an orientation towards socialism". With this new political direction, policies started to embrace a neoliberal economy-oriented spin. Besides prosperity-building, economic growth gained importance. Privatization also returned to Vietnam as the state leadership bid the planned economy farewell. Alongside with decentralization and internationalization efforts under *Đổi Mới* came new values; individuals were re-centred in the new economy as entrepreneurs who now held a greater responsibility as economic actors—in the role of producers and consumers (Schwenkel and Leshkowich 2012). They play a major role in environmental governance as we shall see throughout this book.

Đổi Mới opened up the country to official development assistance (ODA), development organizations and non-governmental organizations (NGOs), and these actors rose in importance in environmental narrative making. Project VIE, for example, is the first evidence of internationalization of environmental policies and its incorporation into the global development efforts. It has been reformed multiple times since, and the latest reform is currently under way. Vietnam's Agenda 21 ("Promulgating the Oriented Strategy for Sustainable Development in Vietnam") was published in 2004 and became another milestone in the country's environmental policymaking. It instigates institutional changes:

> To implement Vietnam Agenda 21 a Sustainable Development Office (Agenda 21 Office) was created located in the Ministry of Planning and Investment (MPI). The office was tasked with devising action programs and plans to facilitate projects in the field of sustainable development. In 2005 the National Council on Sustainable Development (NCSD) was established, tasked with leading the Agenda (Nguyen 2012, p. 21).

The MPI, thereby, manifested its strong role in the hierarchy of ministries responsible for environmental policymaking and Vietnam adopted Sustainable Development to its national context. In 2008, climate change became the focus of policymakers due to the increased efforts of international NGOs, donor organizations and development aid agencies putting the topic up on the agenda (author's own interviews). The NTP-RCC was adopted alongside implementing orders. Climate change has been and continues to be a central topic in Vietnam's environmental governance because of the high risk that this crisis poses to the country. Among the climate change papers introduced after 2008 are Vietnam's Intended Nationally Determined Contributions (INDCs) in 2015, which has been recognized as the most ambitious in the region.

Another crucial addition to the government's environmental regime is the Green Growth Strategy (2012), which was updated in 2022. It strengthens the economic-environmental nexus embedded in the Sustainable Development paradigm. While there is a strong focus on environmental problems, solutions are framed economically under the ecological modernization programme. Still, the Green Growth Strategy is in line with the climate change strategy, and it is coordinated by the National Committee on Climate Change. As one of the measures to achieve this goal, Vietnam has, for example, introduced an environmental protection tax. In addition, it focuses on reducing energy consumption and shifting to renewable energy sources (Quitzow, Baer, and

Jakob 2013, p. 18). Biodiversity and forest management are further priority areas in Vietnam's environmental policy.

From Legislation to Execution: Shortcomings in the Implementation of Environmental Policies

Vietnamese government institutions are complimented for their progressive environmental law-making during interviews with NGO staffs and in academic literature (Schirmbeck 2017; Ortmann 2017). While the government is praised for setting up the legal framework, a lack of implementation strategies exists (Nguyen 2012; author's own interviews). There is a clear gap between policy narrative and implementation that is described by scholars and NGOs alike. Nguyen (2012, p. 21) explains this gap as a lack of capacities and environmental awareness. Mahanty and Dang (2015) trace the lack of capacities to individuals in specific positions in government institutions. Wells-Dang (2011) also notes the importance of the individual actor's personal network, beside the network of the institutions that they work for.

Below the national scale, the ministries have their provincial counterparts whose main responsibility is to adapt the national law to their own context. Mahanty and Dang (2015) find that "local cadres' horizontal linkages (commune authorities to villagers) were stronger than their vertical linkages up to higher levels of government". Ortmann (2020) sees the gap between law and implementation as the missing translation and weak networks between the scales. He describes how the provincial ministries were not penalized for falling short on law implementation. Thus, it is important to understand the power that accumulates at the local scale and how narratives are translatable across scales.

Another reason for the shortcomings in implementation lies in practices of cronyism, favouritism and corruption. Benedikter (2016, p. 26) quotes the idiom "If one becomes a mandarin, the whole lineage asks for favors" to underline how a person's network affects their policymaking. Furthermore, interest groups have risen in importance over the past years and corruption has become a part of the Đổi Mới process (Vasavakul 2019, pp. 1–2). Interest groups, through their relations, managed to position their vested interests in the centre of policymaking within the official state structure. Corruption, therefore, spread from individual government officials taking bribes to making regulations with non-state actors for investment returns.

The high level of bureaucracy and unclear responsibilities throughout the hierarchy are also seen as reasons for the lack of policy implementation. The government has founded several agencies, academies and centres whose tasks are to conduct research and provide consultancy on environmental issues. The Vietnam Academy of Science and Technology is the largest research institute in the country conducting research on environmental concerns. In 2006, the Institute of Strategy and Policy on Natural Resources and Environment (ISPONRE) was established under the MONRE to conduct research and develop strategies for the ministry (Ortmann 2017, p. 78).

Environmental state governance in connection with socioeconomic politics has been an integral part of the Socialist Republic of Vietnam since its establishment. It is used to establish social control while at the same time, social politics and economic growth have been used as the reason for weak environmental regulations. This results in weak environmental policymaking across authorities whose cooperation is limited, and responsibilities are not always clear. While environmental policy in Vietnam looks promising on paper, the implementation is lacking. Other systemic problems, such as lack of capacities and corruption, further weaken policy implementation. Non-actors need to work within this context and find ways to address problems in environmental policymaking.

Against this structural and legal background, NGOs were introduced to environmental work that spans across the interaction between humans and non-human actors. The work involved does not solely focus on human relations. Instead, environmental NGOs acknowledge the importance of human-nature relations, and they seek to incorporate these relations in their work.

The Repressive Side of the State: NGOs and Environmental Action

With the restructuring of political-economic systems and the opening of the country towards international actors, NGOs, civil society, academia and international organizations gained entrance into environmental governance and started to transform environmental policies (Fortier 2010, p. 232). Despite being an authoritarian state, Vietnam, nevertheless, has political leverage. The ancient proverb "the laws of the king are less than the customs of the village" (*phép vua thua lệ làng*) continues to be meaningful today. Surely, the villages themselves are aware of the hierarchy, but they are not equally organized, and their social culture continues to vary throughout Vietnam till the present day. Kerkvliet (2005) states that the role of the villages has been diminished. He argues that land collectivization was backtracked as farmers did not support collective farming and rather left their land barren. Therefore, individual usage rights were reintroduced.

Looking at the historic reach of state power, we can see that the government's handling of protests and critics continues to be embedded in this socio-political bureaucratic structure that also formed the basis for the way NGOs function in Vietnam today. While some protests have been allowed, which results in critical voices being incorporated into the policymaking process, others are harshly suppressed. Those facing repressions are mostly individuals who questioned the legitimacy of the CPV system and the national politics, or they attack the country's vested interests (Ortmann 2021). These individuals are put on trial publicly to warn others. Although there are few such individuals, the state uses much effort to silence such critical voices. Methods of silencing opposition ranges from surveillance and vocal threats to arrests that result in jail sentences and even the death penalty (Pham Thi and Behrens 2019).

The latest arrest that has sustainably changed the NGO-state interaction was that of Nguy Thi Khanh in early 2022. Khanh is the first Vietnamese to win the Goldman Environmental Prize, which is highly regarded by the international environmental community. She was arrested for alleged tax evasion because she did not pay taxes on the prize won. As she has good relations with many government officials (she even travelled as a member of the official Vietnamese delegation to the Glasgow Climate Change Conference in 2021) and her expertise was valued, her arrest sent shock waves throughout the NGO world in Vietnam. Although she has since been released, her case sets the trend of harsher crackdowns on activists by the state.

Protests are usually rare in the authoritarian country, but environmental conflicts and their narrative have been discussed. Nguyen (2020) wrote:

> Environmental citizenship in Vietnam during the past decade has been primarily manifested through environmental activism by groups of Vietnamese citizens, mostly the grassroots, challenging the state's monopoly in environmental politics. Such manifestations range from micro-act of online participation (such as reposting images or post/critical viewpoints from activists, "liking" posts, posting comments and Facebook live videos, signing petitions) to participation in physical protest events. Three modes of activism: activism for environmental justice, activism for ecological justice, activism for both.

Activism not only challenged the state's role in environmental governance, but also questioned the networks between the state and businesses. The legitimacy of the state has been called into question in its framing of the space for its citizens. The people saw that activism went beyond their own agency and it refocused on power relations. For example, protests that occur at the village level are often ignored. Critique on the local level is widespread. Personal entanglements do matter here, and the lines blur between where one's personal networks start and where the official party's networks end. The multiple identities of individual actors lead to criticism of the state's channels (Behrens 2023).

The scale of the protest and the actors that are involved do matter. For example, land-rights protests are hardly featured in the state media, neither do small-scale protests by farmers. In 2020, land-rights protest in Đồng Tâm ended tragically when three police officers and the local village head died during an attempt to displace the local community from an area claimed by the military-owned company Viettel. Another example is environmental protests. In 2015, the protest against the cutting of trees in Hanoi was successful after the mayor of Hanoi agreed to plant new trees elsewhere to replace the felled trees. But, in 2016, when the toxic spillage from the Formosa plant in Central Vietnam resulted in protests that spread beyond Central Vietnam to Hanoi and Ho Chi Minh City and put into question the state's power vis-à-vis the Formosa Corporation, the protests were violently dissolved. In the latter case, not only did the protests move to the national scale, but they also questioned government institutions and the Party's legitimacy and even threatened to replace the Party's network. Similar strategies can also be observed during the protests for ethnic minorities rights and the anti-China protests concerning the South China Sea policy. It should be noted that critique that are expressed within the official apparatus and do not question the state's legitimacy are allowed (O'Rourke 2004; Hannah 2007; Kerkvliet 2019; Ortmann 2021).

NGOs are concerned about when and how they can express criticism safely in an authoritarian system. By definition, they function outside the government, yet to some extent, they need to be aligned to the state. The CPV is supposed to care for its people and organize societal voices within the existing structures. In theory, the people are the leaders of a socialist government and are organized into the party's mass organizations. NGOs, therefore, need to be perceived as supporters and aid to the state. They cannot be seen as opponents, but they must not lose sight of their goals and values. Thus, the functions and roles of NGOs are often contested. The next chapter will discuss this further.

Environmental actions are not limited to NGOs only. Other actors, such as landscapes, citizens, universities and businesses, are needed for a more holistic understanding of the governmental architecture. A later section of this chapter discusses the role of landscapes in policymaking. The last chapter of this book discusses the role of citizens in late socialist market economy. I shall now discuss how universities and businesses formed networks with the NGOs and the state.

Other Actors and Environmental Action

Universities play an important role in developing environmental narratives. Specialized universities and their research centres contribute to discourses on the environment (Zink 2013). Environmental discipline is usually under the faculties of natural sciences in universities and colleges. As Vietnamese education system is based on the Soviet model, there are several universities, such as the Water Resources University in Hanoi, which specialized in environmental research. Due to the Soviet tradition, research is conducted not only at the universities, but also at the affiliated research institutes. Vietnamese policymakers rely on scientific expertise to inform their law-making process (Zink 2013; Weger 2019), and NGOs tend to cooperate with them. Both Vietnamese policymakers and NGOs see the scientization of the environmental narratives as key for their actions, and universities thereby become important in the governmental architecture that encompasses ecological modernization.

Another group of actors that are vital to the ecological modernization programme and the state's embrace of capitalist market economy are private economic entities. As the economy has grown since the 1980s in Vietnam, private economic entities have become an important actor in environmental narratives and practices. The economic system establishes path-dependencies for discourses. Although private enterprises are permitted, state-owned enterprises still play a vital role. In the energy sector, for example, two-thirds of coal power enterprises are owned by the state (Global Energy Monitor 2022) and are thereby targeted for green energy transitions.

Besides the state-owned enterprises, private businesses, which require permits for business activities and access to resources, also want to maintain a good relationship with the state to implement their vested interests. Depending on their size, their ability to create value is their bargaining power and interest to state officials. Generally, the interconnectedness between the government and private business sector are hard to be assessed. What are more obvious are the dual legal frameworks for the institutionalization of state and private businesses. The influence of economic vested interest groups (e.g., private-public partnerships) is relevant when it comes to environmental policymaking and implementation (Wischermann 2011). In many cases, their interests are contradictory to environmental protection. This is seen in the investment by the country's largest business groups, Sun Group and Vingroup, to convert national parks (e.g., Cát Bà National Park, Tam Đảo National Park, Phong Nha area) into places of consumption, tourism and entertainment. In the Formosa scandal mentioned earlier, in 2016, the Taiwanese firm polluted the coast in Central Vietnam, leading to the loss of livelihoods and impacted the health of the local population. The nonchalant response of the government towards the violation of environmental standards consequently sparked nation-wide protests (Fan et al. 2020). Environmental initiatives, nevertheless, exist in practice. Foundations, such as the Coca-Cola Foundation and Kodak Foundation, finance anti-plastic garbage campaigns. Additionally, beyond this antagonistic relationship between the environment and businesses, there have been several "eco-businesses" established over the past years that use environment as their unique selling point and business model. Although ecotourism companies, organic food dealers, bamboo product companies and zero waste shops fill a niche for the middle class and the foreigner community in Vietnam, they diversify the environment narrative and are part of the government's green consumerism initiatives. These companies are large-scale enterprises and not small and medium-sized enterprises (SMEs), which comprise 97 per cent of registered firms. With five million labourers, SMEs contributed 45 per cent to the country's GDP in 2022 (Dung 2023), but little is known about the relationship between environmental governance and SMEs.

Businesses regardless of size actively pursue commodification, technization and privatization of the ecological modernization programme. Whether private enterprises support or threaten the governance of the Vietnamese state is not the subject of this research. Instead, this book touches on NGOs and their role in narrative-making and their environmental practice. The last section of this chapter introduces one more actor—nature—in the development of environmental governance in Vietnam. McElwee's (2016) concept of environmental rule looks beyond top-down policy processes to understand power structures in environmental policies. It looks beyond human actors to look at nature. The following section describes the importance of landscapes and their entanglements with policymaking during the period of communist struggles and leadership.

Nature as Actor: Entanglements between Landscapes and Policymaking in Socialist Vietnam

Nature and the environment have played a role in the formation of the communist state in Vietnam. Even though Vietnam had declared independence in 1945, the French refused to grant independence to their colonies. The communist independence movement entered a war with the colonizers that eventually led to the victory of the Socialist Republic of Vietnam after fighting back Chinese forces. The environment played a major role during the war against the French and the Americans.

In the Mekong Delta, the Việt Minh used the fragile infrastructure in their war strategy. They destroyed floating roads and bridges to restrict French access to the canals (Biggs 2010, p. 139). Beside as a warfare strategy, the destruction of infrastructure and construction of barriers led to the renaturation of plantation land in the delta. The US troops also understood the role of the environment actor and attempted to destroy this actor by chemical warfare. During 1962–71, 40 per cent of the mangroves in South and Central Vietnam were destroyed (Hong and Hoang 1993). Biggs (2010, p. 174) further argues that the environmental policy of Ngô Đình Diệm (president of the Southern Republic of Vietnam) strengthened the Việt Minh in South Vietnam, as his attempts to stabilize the delta through the engineering of dikes, canals and pumping stations left the already devastated area even more fragile. This pushed farmers in the area to support the Việt Minh.

Điện Biên Phủ's geography and its connection with the Việt troops played a major part in the battle against the French forces in 1954. The Vietnamese adjusted to the mountainous battleground by building mobile trenches from the French camp (Goscha 2016, p. 286). General Võ Nguyên Giáp describes in his memoirs how the strategic planning of the landscape helps them to win the battle. He recounts from a conversation with a Czech journalist before the Vietnamese victory in Điện Biên Phủ: "I rarely write poems, but this view is truly poetic. We are about to do battle just to make the whole country as beautiful as it is here tonight" (Võ 2004, p. 19).

The forests in Western Vietnam, Laos and Cambodia where the Hồ Chí Minh trail was established, served as a cover for the communist forces. It was so effective that the US forces used chemical warfare on the forest as part of their war strategy. The spraying of toxins on the forests had a lasting impact on Vietnam's environment as well as on its people (De Koninck 1998, p. 349). Till today, the soil remains poisonous and causes severe disabilities in the people (author's interview).

The role of environment in warfare and the cooperation of the communist forces dominated the North Vietnamese government until the war ended in 1975. Several notable quotes on the environment are attributed to leading communists such as Hồ Chí Minh and Võ Nguyên Giáp. When Hồ Chí Minh declared Cúc Phương as North Vietnam's first protected area in 1962, he said, "Forests are gold; if we know how to protect and develop them well, they will be very precious" (McElwee 2016, p. 3). In the early 1980s, Võ Nguyên Giáp said, "The soldier comes to another front now, the environmental front … without environmental recovery, Vietnam cannot have economic recovery" (cit. in Beresford and Fraser 1992). The Vietnamese National Archives III, the earliest government record of official environmental policy that dates to 1972, underlines the importance of this nexus for early communist rule. Another surge of environmental topics came in the second half of the 1980s. Environment has always been a concern of the communist government—at times even cutting across other policy areas and connecting to various narratives of war, health, resources, and intrinsic as well as extrinsic values. Intrinsic values are affective factors stemming from human's inner emotional landscape, for example love and health; extrinsic values include external, systemic factors that shape value sets, like financial wealth.

Due to the ongoing war, conservation and environmental governance were topics of interest to the government of Northern Vietnam in the early 1960s. In 1961, the Forest Inventory and Planning Institute was established and a year later, the General Department of Forestry was founded, along with the forest reserve Cúc Phương, which would later become a national park (Déry and Vanhooren 2011, p. 191). Vietnam subsequently established numerous protected areas of various sizes, which led to a "fragmented" landscape of protected zones; by 1969, there were twelve such areas (Déry and Vanhooren 2011, pp. 193–94). Due to the war and its consequences, the expansion of national parks and other areas was halted until after the renovation politics of 1986, when a new effort was put forth establishing another 73 protected areas (Déry and Vanhooren 2011, p. 196). The purpose of declaring protected areas was not only to solve ecological problems, but also serves as a form of environmental rule that would become an important cornerstone for biopolitics

(McElwee 2016). Controlling the protected areas meant controlling the people who were living around the areas. Through policies on the protected areas, the government could interfere with what people consume (e.g., disallowing the destruction of certain species), what they traded and what agricultural practices were allowed. In some cases, this meant an interference with the lifestyles of semi-nomadic communities or forest dwellers.

With the victory of communist forces in North Vietnam in 1954 and the reunification of Vietnam under the communist regime in 1975, a new environmental political framework was established under the official state ideology of Marxism-Leninism. Whether Marxism is ecological has been debated in academia for decades. On the one side, critics of an ecological Marxism point to the lack of consideration for the environment and the exploitation of resources. Marx regarded materialism as enhancing the commoditization of natural resources and disregarding the creation of surplus value from environmental sources (Callicott 1989). Even in socialism, industrialization requires inputs from the environment (Saed 2019). Additionally, Marx dismisses many ways of understanding the separation between human-nature relations and economic concerns.

On the other side, scholars have expanded on Marxist concepts and adapted them to current challenges. Foster (1999) developed a string of ecological Marxism using the concept of metabolism between humans and the environment. He expanded on Marx's metabolic theory to understand the environmental crisis in capitalism (Lievens n.d., p. 6). Skirbekk (1994, p. 99) suggests applying the theory of surplus value to understand extractive economies. Kohei Saito (2023) even urges us to rethink Marx's theoretical considerations as "degrowth communism".

Marxism-Leninism allows the government to interpret and define environmental policy within this ideological framework. Looking at practices in Vietnam since 1975, government policies are centred around developing prosperity with a focus on industrialization. Environment is part of the policy but not a deciding factor. The three waves of agrarian and capitalist reforms during 1953–56, 1958–60, 1976–78 as defined by Vũ Tường (2017) were characterized by different political foci, which altered the relationship between people and their land, as well as power relations. During the third wave, the focus was on rebuilding the country and developing infrastructure projects that required an expansion of industrial forces in the country. The then prime minister Phạm Văn Đồng proclaimed rapid economic growth as a major goal, which was also underlined in the five-year plan (Schwenkel 2020, p. 109). This was understood to mean developing technology for modernization. These efforts intensified after income growth only reached 0.4 per cent between 1975 and 1980, and output in state-controlled industries sunk by 6.5 per cent, despite the ambitious goals (Porter 1993, p. 50). In the following decade, economic numbers rose but it was during *Đổi Mới* that Vietnam's socioeconomic plan improved, and the market-based capitalist ideology was manifested as part of Vietnam's version of a Marxist-Leninist Socialist state.

In agriculture, the government thrived towards a more productive and modern farming sector. Multiple reforms were introduced. Private land ownership was first abolished, and agricultural production was managed by state cooperatives that co-existed in a "hybrid complex organization and property rights regimes" (Adger, Kelly, and Nguyen 2001, p. 79; Spoor, Heerink, and Qu 2007, p. 20). A modernized agricultural method was implemented on the collectivized land that, according to the Soviet model, saw growing prosperity as a main concern (Skirbekk 1994, p. 96). Technology helps to pave the way for a growth doctrine in agricultural production, similar to the industrialization that happened in urban areas (Scott 1998).

Another reform in land politics was executed by the Communist Party, whereby Kinh farmers from the Red River Delta were resettled in highlands and tasked to introduce wet rice agriculture in the lands of ethnic minorities (Lundberg 2004, p. 55). This led to a new wave of deforestation similar to establishing protected forests elsewhere in the country. Shifting cultivation was enabled by redefining the forests and their purpose. The reform was implemented based on French influence, traditional Kinh agricultural practices and Communist government plans that perceived the mountainous regions of Vietnam as "remote, dangerous, sparsely populated and backward part of the country, and with a population that more or less have stagnated in development" (Lundberg 2004, p. 37). Resettlement became necessary as the agricultural areas in the Kinh land seemed to have been greatly exploited since the early 1960s (Lundberg 2004, p. 47). The resettled Kinh were reluctant to change their habits and culture due to their perceived superiority over the ethnic minorities, as well as the demand for their produce from the rest of the country (Lundberg 2004, p. 129). Déry (2000, p. 36) identifies this process as a continuation of colonization in the sense that an external form of governance was forced upon formerly autonomous areas.

The delta land has been encouraging settlement and wet rice cultivation, whereas the mountainous regions, home to ethnic minorities, have enabled shifting and swidden cultivation. This difference is reflected in discriminatory practices till today: for example, ethnic minorities were portrayed as backward and were forced to change their way of cultivation.

In the land of the Kinh people, the preference for a landscape that is useful to humans correlates with agricultural techniques: "In peasants' views in general, as well as in perceptions of government officials, wilderness and wildness are often closely connected. People living in an undomesticated nature are almost by definition 'wild,' and uncivilized people" (Persoon 1997, p. 2). Agriculture, thereby, becomes a practice through which we understand environmental narratives and power relations. An "orderly" agriculture provides taxes that are paid to the central authority and contributes to the government plans and policies.

In contrast to this centralized network of villages in the lowlands, upland societies organized themselves in a more decentralized and fluid manner. This can be attributed to the inaccessible geography and the shifting agriculture in the forest uplands, contributing to the resistance against the centralized state and variations in environmental narratives. Ethnic minorities continue to be seen by the Kinh majority as uncivilized and wild. Following the notion that "nature" become better when cultivated for human needs, planted fields are seen as a sign of civilization (Lundberg 2004, p. 141). Interestingly, this perception seems to shift with rapid urbanization. Today, on the one hand, ethnic minority agricultural practices are still blamed for the destruction of nature, but on the other hand, ethnic minorities are romanticized for being close to nature and having a pure relationship with it.

Another dichotomy pair characterizing this process is "*lạc hậu*" and "*văn minh*", which basically means "backward" and "developed/progressive", respectively. In literal terms, "*văn minh*" means enlightened, educated or literate and underlines the importance of education. Vietnamese often use these words instead of "*phát triển*" (developed). Rumsby (2020) shows how Hmong communities use the term "*văn minh*" to internally create divides between communities that have converted to Christianity and those that have not. Similarly, ethnic minorities use these phrases to define their own meaning of development.

The history of the term "*văn minh*" shows the cultural transboundary background of Vietnam and the nationalization of foreign terms in the Vietnamese context. This term derives from the Chinese word *wenming* (文明). In the Chinese context, *wenming* was a crucial principle in the tribute trade system of Chinese foreign politics, describing a way of enlightenment that was tied to divine behaviours of heaven, earth and people (Chang 2016, p. 5). The term stands for sacredness and is used to underline China's higher position and development status compared to their surrounding countries. In Vietnam, the word became important when a translation was needed for the French term civilization, a concept imported through colonialism. A Vietnamese reform movement (*Duy Tân*) appropriated the word to call for reforms that resembled the Japanese Meiji Restoration, postulating Western civilization as a model but at the same time paying tribute to the local culture (Chang 2016, p. 1). Several writers picked up the term to discuss the state of society, and even Emperor Khải Định argued that monarchy was an expression of the term (Chang 2016, p. 2). The word continues to thrive today with a history of Chinese influences, describing French civilization with respect to Vietnamese sociopolitical systems. It continues to express a space of debate in which people create a perceived hierarchy within society. This hierarchical thinking and othering make it similar to the idea behind development.

Social norms have shaped the way humans interact with the environment, but environmental conditions have informed governance and power structures. The centralized government system was necessary to manage the canal and dike system of the Red River Delta to ensure sufficient rice production for the kingdom. However, villages and individual families were responsible for rice farming and the food supply. This structure has reemerged after the failed land reforms in the 1950s (Kerkvliet 2005). Consequently, the formation of the Vietnamese state continued to rely on centralized power structures, with autonomy given for smaller units and a party structure reaching into villages.

Urban, peri-urban and rural livelihoods influence people's and organizations' practices and views on nature. Not only have people and institutions created cities and rural areas, but these spaces have also influenced and shaped the mindsets of the people living there. Adger, Kelly, and Nguyen (2001, p. 4) observed that "the relationship between the rural and urban economics of Vietnam in the early part of the transition and the more recent reforms is crucial in explaining the nature of the present-day economic structure and institutions". Fuhrmann (2017) has shown in her research how the peri-urban environment in the Mekong Delta influences the perception of environmental change, and due to its locality is shaped by its relationship to water and agriculture, and by concepts of modern urbanism. Therefore, the locality of projects and their landscape need to be taken into consideration when analysing the case study in Chapter 5.

The process of urbanization continued under the communist government and was interrelated to modernization and industrialization. Rapid urban growth has consequences on the environment, but it also isolated humans from nature. Schwenkel (2017, p. 45) shows how the planning of cities and towns followed "urban eco-socialist ideals". Such planning emphasized green spaces in urban areas for health reasons. It limits pollution and supports the mental health of workers, who could use the space for recreational means. Hence, nature was converted into a space for enjoyment (Schwenkel 2017). This notion was challenged, however, when food scarcity led to the establishment of urban agriculture. Furthermore, building houses was ecological as they were built for living with limited resources (Schwenkel 2017, p. 53). The narrative cultivated through this urban eco-socialist architecture was, nevertheless, one of pairing the enjoyment of nature with using it for livelihoods.

In an article published on the 131st birthday of Hồ Chí Minh and the elections for the 15th National Assembly in May 2021, General Secretary Nguyễn Phú Trọng discussed "a number of theoretical and practical issues on socialism and the path towards socialism in Vietnam" (Nguyễn Phú Trọng 2021). In this article, Trọng wrote that "we need sustainable development in harmony with nature to secure a clean-living environment for present and future generations, instead of unlimited exploitation and possession of resources, unrestrained consumption and destruction of the environment" (Nguyễn Phú Trọng 2021) and called environmental protection "an existential issue". The government recognizes environmental concerns as central to Vietnamese policymaking.

Landscapes become actants in ruling specific areas of the country. For example, tourism as a source of revenue has grown rapidly over the last decade. Beyond the famous Ha Long Bay and Sa Pa, other destinations like Ha Giang, Pu Luong, Phong Nha and Phu Quoc also attract both domestic and international tourism and shape the provinces' economic strategies. Those provinces with coal reserves or limestone mountains exploit these natural resources, while other provinces that experience a lot of sunshine and wind become the centres for energy transition towards renewable energy in Vietnam. Landscapes, thereby, inform political-economic decision-making and continue to be altered as limestone disappears, rivers turn into dams and wind power parks create a new landscape of modernity that informed our understanding of sustainable development.

Conclusion

This chapter outlines the context needed to understand the role of sustainable development in Vietnam. It establishes a basic understanding of relevant institutions in environmental policymaking and the important divergence between environmental actions in theory and the actual implementation of laws and decrees. The state actors' relations to society have been contextualized. Other relevant actors in environmental governance in Vietnam were also introduced. Among them is landscapes that exemplify how nature has influenced political developments in Socialist Vietnam. This background helps to understand the complex interrelations of institutionalized and non-institutionalized actors and systems that affect environmental rule. With this context in mind, the next chapter discusses the conceptualization of NGOs and civil society in the Vietnamese context to better understand state-society relations and the actors involved in environmental rule.

3

NGOs and CSOs in the Vietnamese Context

The Academic Debate on What Makes an Organization Non-governmental

Whether non-governmental organizations (NGOs) and civil society organizations (CSOs) exist in Vietnam's political system, how they earned their names and what is their positionality, have been thoroughly debated among Vietnam studies scholars. This section contributes to this debate by providing recent data on the self-identification of NGOs, their roles and their positioning in networks. It shows that although none of the organizations are 100 per cent non-governmental, they still perceive themselves and are perceived as separated from the state and build their networks accordingly. These organizations may vary in their goals, but they nevertheless show similarities in their approaches and networks due to the limited opportunities in Vietnam's political context and the donor landscape. The restrictive authoritarian and "responsive-repressive" system forms the background for actors in the wider Vietnamese political sphere and casts doubts as to whether NGOs and civil society (CS) exist in Vietnam. The definition of an NGO in this research is broader than the literal meaning of "non-governmental organization". The terms NGOs, CS and CSOs can overlap, complement or contradict each other. To reach a common understanding of these terms, I start with deconstructing the term NGO before turning to the concept of CS and discuss what these concepts mean for Vietnam.

The term NGO was first used in the UN Charter of 1945, when NGOs were accepted as unofficial consultants and were explicitly named as such. This established a dichotomy between the official government actors and the unofficial non-government ones that were characterized by some form of organizational structure and closed identity. Since then, an academic debate has unfolded on what constitutes an NGO, besides being non-governmental. Vasavakul (2003, p. 25) notes that the Vietnamese political sphere refers to NGOs at the end of the 1990s as "popular organizations that were loosely connected to the party-state structure, financially self-sufficient and involved in development work once considered as falling under the jurisdiction of the party-state". Vietnamese NGOs (VNGOs) used the methodology of fence-breaking (*phá rào*), which meant that policy positions were not directly expressed, but indirectly through breaking rules and regulations, which is still happening today. In 1992, under Decree 35/CP the government allowed the formation of private, non-profit social organizations, which still form the basis for various forms of organizations today. In 1994, the United States ended its embargo on Vietnam and consequently, more international organizations entered the country.

Before proceeding to define NGOs, it is worth stating what they are not. What qualifies as not belonging to the state's government? Several scholars rightly argue that NGOs today have an active role in governance, and they are in one way or another connected to states, not only in authoritarian settings like Vietnam, but also in democracies where several of them rely on government funding. NGOs are globally connected via networks and co-dependent on government institutions. Still, they can initiate actions independently from government actors; they form part of governance but are not government actors. This can lead to NGOs becoming agents of resistance and be subversive to state power (Princen and Finger 1994). They can strengthen institutionalized as well as non-institutionalized resistance by providing networks, access to information and other resources. Their capacity can be used by other CS actors, and these coalitions are able to apply pressure on the governance system where single actors might have failed to do so.

Regarding their positioning, NGOs can position themselves in both bottom-up and top-down policymaking. On the one hand, democracy provides significant (if not necessary) political conditions for the existence and activities of NGOs. At the same time, NGOs can enhance the quality of democracy. As pressure groups, NGOs "permit citizens to express their views on complex issues which affect their lives", which is important because "democracy cannot be simply reduced to a head-counting exercise: it must also take account of the strength of feelings expressed, and of the quality of arguments advanced" (Grant 2000, pp. 35–36).

On the other hand, scholars argue that NGOs are formed to advocate a public policy (Petras 1999). They seek to change but not to fundamentally reform the system. They are a new ruling class because they have capacity to control the vulnerable and have access to financial resources due to their organizational structure. They do not produce commodity but instead act as a service provider. The question is who receives their service: the target group of their projects, the donors or the government. Therefore, NGOs can become compliant with government policies and (un)intendedly support them and strengthen state structures. Especially in authoritarian contexts, they can easily become a supporter of state structures when manoeuvring their work within it. In China, Weller (2006, pp. 123–24) made the following conclusion, which is also applicable to the Vietnamese context:

> Under corporatist arrangements, NGOs either accommodate to the state or find themselves dismantled. Those that survive tend to have the centralized structures that authoritarian states prefer to deal with. They also look modern in the sense that they are strongly institutionalized, with clear bureaucratic mechanisms, dealing with issues that these self-consciously modernizing states find appropriate. Neither local community temples nor lineages, for example, have these features. The result in many authoritarian regimes with roughly corporatist relations to society is a bifurcation between local, informal associations that survive beyond the gaze of the state, and these centralized, formally organized NGOs that survive by working closely with the state.

This organizational structure and networks that include either state institutions or personnel can be found in the case studies throughout this book. By cooperating with state actors and taking over responsibilities from government institutions, NGOs add to the legitimization and stabilization of the authoritarian state power in Vietnam. At the same time, they translate resistance into policy advocacy to achieve policy change.

From the perspective of Area Studies, narratives and concepts are specific to each context and every individual occupies a unique position (Cho, Crenshaw, and McCall 2013). The morality of what constitutes "good" and "democratic", therefore, must be revaluated for every case. In this research, morality is a self-perceived definition by NGOs, which they practise with different strategies. This variety of approaches and positioning of the NGOs can result in denying another organization the status of an NGO, according to moral standards. Because morality is crucial for the governing of the Vietnamese state in the environmental sector, NGOs need to prove that they are morally committed to the political-economic system. According to Derks and Nguyen (2020, p. 2), development has taken a moral turn and "political-economic matters are reframed as moral imperatives". This means that the responsibility for development success is no longer solely dependent on the state and large development organizations. Instead, success depends on "moral qualities" of individuals, households or communities—or even NGOs. In line with the individualization and decentralization of a market-based governance, this re-shifting of responsibility avoids asking questions of "inequality, domination and injustice" (Derks and Nguyen 2020). Morality has, therefore, become an overall expectation towards actors in the environmental field.

Moving away from morality and towards a structural definition of the relationship between NGOs and the state in Vietnam, Wong (2012) argues that NGOs organize themselves according to their access to power. She defines NGOs according to their organizational structure in relation to their work effectivity and how they shape norms. Therefore, the state has an indirect influence on the NGOs' organizational structure and their agenda (Wong 2012). An example is ActionAid International that legally founded two sister organizations to be able to fully function under Vietnamese law. Most organizations have separate fields of work to cooperate with government institutions and with communities. The NGOs themselves perceive the multi-level strategy as an asset in strategic advocacy.

Yasuda (2015, p. 24) defines NGOs as non-profit organizations that advocate for human development. However, the concept of non-profit is difficult to concretize. In times when NGOs have difficulty obtaining permits for their activities, they transform into social enterprise (author's interviews). They even perform consultancy work at times to financially support their endeavours. The non-profit trait, therefore, is ambiguous as there is no universal understanding of what constitutes profit-making. How much must an

NGO pay its staff or its managing board or invest in its operations without being pro-profit?

In many countries, there are legal frameworks in place for this issue, resulting in different standards in the international networks. In Vietnam, the legal framework consists of Decree 12/2012/ND-CP on the registration and administration of foreign non-governmental organizations and Decree 93/2019/ND-CP on organization and operation of social funds and charity funds. Non-profit is therein defined as "The profit earned during the operation is not to be divided but only used for activities in accordance with the recognized fund's charter"[1] (Government of the Socialist Republic of Vietnam 2019). Activities may include supporting the development of culture, education, health, sports, science, agriculture, rural development and in cases of natural disasters or similar incidents.

During my interviews, NGOs accused each other of not being proper because they either make too much money or seem to be doing projects for the sake of getting financial grants, and not for a moral reason. Intransparency, widespread corruption and missing accountability made the problem worse as it is often not clear if organizations make profit or not (Gainsborough 2010; Vasavakul 2019).

The roles of NGOs are as diverse as their adaption approaches. Doherty and Doyle (2018) recognize these differences between NGOs in how they position themselves as non-profit, non-governmental and on their morality levels. They are divided into governance groups or emancipatory groups according to their environmental narratives. Governance groups are embedded in the neoliberal framework as they do not address environmental justice. Emancipatory groups, on the other hand, are, at least on a discourse level, critical of power structures and seek to create networks across scales and spaces for alternative narratives and new actors to emerge. There are organizations in Vietnam that challenge the state's governance, but they also then experience oppressive actions of the state. Thus, it becomes essential for NGOs to navigate the fine line between what they can and cannot do, become governance actors and be in line with governance and state-set narratives.

---

1. *"Là lợi nhuận có được trong quá trình hoạt động không để phân chia mà chỉ dùng cho các hoạt động theo điều lệ của quỹ đã được công nhận."*

Definitions of NGOs from the State's Point of View

Most VNGOs are registered under the Vietnam Union of Science and Technology Associations (VUSTA). VUSTA has previously identified itself as an NGO (Vasavakul 2003, p. 43), but it is actually an umbrella organization under the Ministry of Science and Technology. Membership of VUSTA is not voluntary, but it enables work to be done legally. Once registered, member organizations must follow decrees implemented. Rules and regulations have become stricter over the last decade, not only for organizations under VUSTA, but also for the umbrella organization itself as well as for other government organizations. Detailed reports must be submitted on implemented activities and organizations must seek permission for events and publications months in advance. Denials of these permits can occur occasionally and especially before major political events, such as the Party Congress. Nevertheless, VUSTA sees itself not so much as a control body and more of a support organ that enables organizations to work together. Article 5 of the VUSTA's charter lists the organization's functions as follows:

1. Gathering and uniting Vietnamese science and technology intellectuals inside and outside the country, coordinating and guiding the operations of member associations.
2. Acting as a bridge between member associations with the bodies of the Party, the State and the Vietnamese Fatherland Front and other organizations to tackle common issues for the VUSTA intellectuals.
3. Representing and protecting the legal rights and interests of its members, member associations, and Vietnamese intellectuals in science and technology. (VUSTA 2018)

With the right partner, VUSTA can indeed advocate for policy change. For example, VUSTA and VUFO (Vietnam Union of Friendship Organizations) were the organizers and hosts for the ASEAN People's Forum during Vietnam's Chairmanship of ASEAN in 2020. They allowed civil society from across ASEAN to have critical discourse and promoted VNGOs' interests by putting them on the agenda of the ASEAN Summit.

International NGOs (INGOs) are registered under the VUFO and must obtain permits for their activities. The number of registered organizations has been constantly on the rise since 1992 when the government issued a decree to allow the formation of private and non-profit social organizations. VUFO is the standing agency of the Committee for Foreign NGO Affairs

(COMINGO) and oversees "people-to-people diplomacy", which means coordinating international activities beyond state diplomacy. The executive body of VUFO is the People's Aid Coordinating Committee (PACCOM). As of 2012, over 900 international organizations have been registered under PACCOM. Interestingly, PACCOM has been reluctant to reveal its current numbers of registered organizations and the amount of financial flow it coordinates. PACCOM understands its task to include fostering international partnerships and assisting in establishing connections between INGOs and VNGOs. Practically, it is also the surveillance body for INGOs, overseeing matters of registration and permits for cooperation, including those for events and publications. Besides their connection to VUFO, many INGOs also cooperate with Vietnamese ministries or other state bodies, depending on their thematic focus.

As a member organization of the Vietnamese Fatherland Front, VUFO blurs the divide between state and civil society and makes the network more complex. The Communist Party of Vietnam (CPV)'s mass organizations play a critical role in the political system of Vietnam. They are supposed to represent people of different interests and integrate them into the state. These organizations often serve as the starting or ending point of one's political career, connects the personnel between the organizations and forms the interlink between party organs. On a national scale, they are more likely to promote party interests than the other way around.

Academic literature commonly characterizes NGOs as non-profit and part of the wider social field (Sikkink 1993; Wong 2012). But even this notion turns out to be contested in Vietnam, as several interviewees have raised concerns about the profit-oriented working attitude of NGOs. Consequently Wong (2012, p. 3) finds it helpful to define NGOs in terms of their structural traits to avoid assumptions and biases. In this research, the term NGO is used as this is how the informants define and perceive themselves, despite their structural overlaps with government institutions. NGOs are defined as organizations that are registered under VUSTA or VUFO.

Being "NGO": Self-identification from NGO Staff

Inside the institutions are individuals who co-shape their organizations. Thus, to better understand NGOs, it is worth examining the people working for them. In terms of their interaction with NGOs, interviewees define their role as educators, listeners, supporters of communities, leaders, documenters of good practices, builders of trust and consensus, and builders of strategic advocacy. To them, NGOs should "inspire, instead of repeating problems" (author's interviews). These broad tasks can be summarized as an in-between actor that translates and connects people and knowledge.

According to the interviews, most people working for NGOs do not necessarily have an education that relates to their job duties. Most of them have an academic background in foreign languages, mainly English, but there are also others with degrees in economy, chemistry, biology, journalism and many more. Compared to those lower in the hierarchy, NGO workers who are higher up in the hierarchy are more likely to have been "internationalized" with a master's degree, or at least undergone an internship abroad. One associate (author's interviews) remarked that the same happens in the ministries; the people do not necessarily have an educational background that fits with their job description, but international education is perceived as a positive asset.

The more experienced NGO workers mentioned a change in motivation to work for NGOs. While in the 1990s and 2000s, salary was the main incentive to work for NGOs, this is not so the case now. As salaries in Vietnam generally increased, NGOs are now rather seen as a good working environment due to perceived lower hierarchies and a less competitive working environment. Working for a good cause is also mentioned as a motivation to work for NGOs. This could potentially be due to the moral turn of development that Derks and Nguyen (2020) have described. Still, NGOs are by far free from hierarchy. The director holds a high degree of power in decision-making and agenda-setting, along with a high level of personal responsibility.

Personal relationships were also mentioned mostly by staff without prior experience in their field of work as the reason for working for an NGO. It seems normal to recruit friends and relatives as personal connections are important in establishing a basic level of trust with a potential employee. Others mentioned a biographical shift in their lives when they realized that they wanted to work for a good cause. For example, an employee of a local NGO recounted how she was working for a mining company when she saw the destruction that mining did to the environment and communities, so she decided to work for environmental projects instead. Most of the people interviewed have already worked for more than one NGO, which gives them a personal network of people that exists parallel to institutionalized networks and working groups.

All the VNGO employees interviewed were from North Vietnam, except one from Central Vietnam. Some organizations have an office in the South, but otherwise, Northern Vietnamese staff are responsible for projects across the country, creating a bias towards enhancing Northern sociocultural and political understandings in areas of the country that historically have different contexts and different landscapes.

According to NGO employees, the urban-rural divide matters when it comes to working with rural communities. Employees who themselves come from the countryside are often entrusted with working in the local communities, because they know the way of communication with the people and understand local livelihoods better. They are tasked with translation of NGO work to the local contexts. Most NGOs are perceived as more of an urban elite that needs the local gatekeepers for translation work and for understanding sociocultural issues. However, not all NGOs see this gap; as a result, projects that were conceptualized in an urban middle-class environment are then implemented in a poverty-ridden rural context. Few NGOs include the target groups and rural communities during the project planning and conceptualization phase. Additionally, these urban middle-class personalities often have a kinship or personal relation with the party officials, which is an important prerogative for receiving permits. In the state-society relations, these networks with the communities and with party officials show that the state-society relations are a negotiation process that happens beyond these two categories.

Apart from the organizational structure and the individuals involved in NGOs, networks are the third category that helps us to understand the positioning of NGOs in Vietnam. With whom NGOs align themselves is a sign of coalition building and boundary setting within environmental governance. The networks are high in complexity. A good example is the study by Pham et al. (2010) on a project regarding environmental services. They find that the government, local organizations, international and Vietnamese NGOs, international agencies and consulting firms have contradictory or overlapping roles in the realization of the process. Capacity building in the form of trainings and workshops are conducted by NGOs and government bodies on different levels. International groups conduct research and support local authorities and VNGOs with diametrically opposed interests. Community-based organizations (CBOs) provide relationships and networks for which INGOs, private companies and government agencies would then compete. In most projects, several actors and intermediaries work together yet at the same time compete against each other, while following their own interests. The degree of involvement of the actors and who "wins" the project and makes narratives depend on the specific power relations in each case.

NGOs differ in how they define their desired change and how far this change should occur. There are NGOs that perceive themselves as a service provider for the government, but lacks funding, knowledge and capacity to develop strategies. The change desired can be very topic specific (e.g., single-use plastic) and can only be achieved under the current systemic circumstances. Other NGOs seek to advocate for policy reform and implementation; for example, shifting Vietnam's power plan from being reliant on fossil fuels to renewable energy, which is essentially seeking a complete change in government strategy to understand power grids as a form of decentralization. This is done within the current governance system with generally institutionalized power relations staying untouched, even if community participation (e.g., mass organizations and the local party chapters) is an element of the projects. International organizations can be found in both mass organizations and local party chapters.

NGOs who seek to overthrow the government or question the legitimacy of the CPV are not allowed to operate. Still, there are a few local organizations that seek to change power relations while leaving the one-party state rhetorically untouched. They want to alter power relations by working on land rights, access and ownership of forests and offering a platform for different environmental narratives to be implemented bottom-up and not top-down. These NGOs are usually organizations that have problems with their legal registration and acquiring permits for their operations. This will be discussed in a case study in the next chapter.

The types of projects and activities that NGOs implement to pursue their thematic goals and their theory of change can be clustered into policy advocacy (e.g., compiling studies and holding workshops and conferences), networking (e.g., holding networking sessions and participating in institutionalized networks), research (e.g., conducting external and internal studies), capacity building (e.g. holding trainings and workshops), community support (e.g., providing knowledge through informal talks and formal workshops) and education (e.g., conducting workshops, producing publications and designing curricula). The focus on one or several of these approaches reflects how NGOs perceive themselves, their role within society and their responsibilities. For example, the Center for Development of Community Initiative and Environment

(C&E) widely uses the education approach in its projects, not merely seeing themselves as a political actor involved in changing the system. INGOs, like WWF, are strongly involved in network building and perceive themselves as translators and managers.

NGOs are credited with strengthening civil society by supporting communities and the democratization process. This is in line with the self-perception of NGOs in Vietnam. For example, NGOs contribute to "societal transformation through community development" (Princen and Finger 1994) by looking at model projects and research that advocate for decentralized renewable energies or waste management. Due to the persistent top-down structures in NGO work as well as the constraints faced by NGOs during project implementation, this role is in reality rather limited. In the case studies in this book, communities voiced concerns that are not necessarily in line with NGO projects, and they do not seem to understand that projects are an opportunity for them to discuss topics and propose solutions. The questions remain: how many of the projects are decided by NGOs themselves and what kind of transformation and development are we seeking in an international and authoritarian context? In interviews with NGO staff, international environmental regimes and narratives, as well as institutionalized governmental framework, remain a work-in-progress.

"NGOs are not connected to communities because of [the] nature of our country", said one NGO representative during our interview. Although the majority of VNGOs and some INGOs work with communities (the latter usually cooperates with local authorities or local NGOs), the connection varies widely among organizations. Overall, top-down approaches still matter. Strategies are formulated in offices in Hanoi in accordance with the visions, head office guidelines or donor calls and regulations. Projects, which are also mostly conceptualized in Hanoi, are then fitted into this overall framework. Most NGOs value community knowledge, but it is unclear how far this knowledge is actually implemented in the projects. All three case studies in this book describe different ways of translating work between different scales of knowledge. In an interesting observation, Hakkarainen (2012) notes that although participation is a crucial concept in most NGO projects, what matters is the participation defined by the project implementers. As the project implementers enter communities with their ideas on what participation means, communities have developed an aversion against this kind of participation, which makes them feel forced, and even risky when community and government representatives are brought together without any safeguard mechanisms in place. Interviews regarding environmental knowledge and narratives suggest similar observations.

Lying in the ambiguous spectrum between state and non-state actors are environmental centres and institutes. Not only are they blurry in their definition of being connected to the state, but some of them consider themselves to be NGOs, while others are research centres and think tanks. What these actors have in common is the term "*trung tâm*" (literally translated to "centre") in their name. O'Rourke (2004, p. 196) characterizes them as follows: "However, while they operate like NGOs, these groups are often associated with government universities or institutes. These organizations have autonomy over individual projects, assuming they can raise funding, but they are constrained in what they can critique or recommend. All of these groups are careful to avoid being overly critical of government policies or programs." Some of them have developed into more autonomous units and changed their target groups and self-perception over the years to become NGOs now. One factor in this change is their growing network with international organizations and increased funding from donors (McHale 2004). But again, they are not a homogenous group. The *trung tâm*s embody much more just working in an authoritarian context and jumping between scales. Self-definition matters here and the centres are not defined in their own category. Instead, their structural commonality with NGOs incorporates them into the NGO definition, provided their self-understanding and networks conform to NGO characteristics. The case study in Chapter 4 of this book looks closer at one of the centres that identifies itself as a VNGO. The next session defines the characteristics of NGO networks when considering some centres as NGOs.

NGO Networks

The characteristic of NGO networks is important insofar as it offers a better understanding of how NGOs identify other organizations as non-NGOs. In their differentiation effort, NGOs engage with other institutionalized actors either as an ally, or as another cooperation partner with their own specified positionality in the Vietnamese context. Although networks are not uniform between all NGOs, what NGOs have in common is their reaffirmation of their positioning as NGOs. Generally, partners are chosen to achieve a strategic goal. Cooperation partners, for example, include institutions of higher education.

Universities are used as partners for education projects and to reach young people as a target group, or if the generation of new knowledge is crucial. This is illustrated in a case study in Chapter 4. Cooperation with mass organizations of the CPV and local initiatives are used as an entry point to communities without having to involve state actors. Ties with businesses can be activated for economic empowerment or fundraising purposes. All actors fill in gaps that the NGOs in their perceived position cannot fill themselves.

Among NGOs themselves, there are formal ways of connecting among INGOs and local NGOs. Working groups have been established by VUSTA as well as by NGOs themselves, and the UN holds regular network meetings. An example of a working group is the Climate Change Working Group (CCWG) that has a secretariat funded by INGOs and donor organizations. These paid positions are cited as the reason why the CCWG has such high visibility and output. The CCWG organizes workshops, conferences and produces publications, mainly with the aim of policy advocacy. It has twelve core member organizations and over 100 affiliated member organizations. Member organizations are both VNGOs and INGOs and the Chair is rotated among them. The CCWG seems to be the only "organization" that explicitly works under the narrative of climate justice and brings together several different narrative approaches, besides sustainability. In contrast to CCWG, government-lead networks are perceived as not useful and another bureaucratic burden. Since coalition building may potentially generate countermovement, the state seeks to maintain control over it. Networks consequently become a surveillance space in which it offers a limited capacity to rediscuss state-defined concepts.

Most NGOs nevertheless stay connected for knowledge exchange and strategies to counter government bureaucracy. An exception is the NGO in case study 3 of Chapter 4. According to this NGO, connecting with other NGOs has not proven useful, and cooperation with other VNGOs has not added value nor helped them with their agenda. In contrast, this NGO even finds it counterproductive as it took up precious staff time. This NGO's theory of change is indeed quite different to other NGOs' because they follow a bottom-up approach that focuses on the communities they are working with. They also do not cooperate with governmental institutions, but with specific individuals in different institutions whom they have built a relationship with over the years. This NGO does establish international networks, however, among communities within Southeast Asia, and with its mostly European donor organizations.

At which level NGOs have their crucial networks established and which are the reference points from where their work is planned and implemented give NGOs their distinction. By virtue of their organizational structure, INGOs, such as WWF, CARE and ActionAid, have a strong international network as their head offices reside on other continents. This is not to say that the centre of power is always with the head offices. WWF, for example, has a director who is sent to the country office. There are content guidelines from their "donor" office and certain funds provided by the head office can only be spent on certain projects. For ActionAid Vietnam, decision-making power and agenda setting are more nationalized. It has a Vietnamese country director and has more power to set its own goals. Nevertheless, discursive and practice frames, for example regarding donation, continue to be set by the international ActionAiders. Interestingly, the donation department is not managed by the country office but under the direct management of ActionAid International. The rest of the operations are within the responsibility of the country director, who, together with the team, decides the application of international directives. In contrast to WWF, ActionAid Vietnam has links to other international organizations (e.g., Bread for the World and the Rosa Luxemburg Foundation) for financial support. The position of an NGO, whether it is defined as an INGO or a VNGO, and the scale of power within the respective organization, is therefore differentiated between the organizations, widening the spectrum between INGOs and VNGOs.

VNGOs built international networks that are used mainly to exchange knowledge and to create spaces that are absent in the Vietnamese sphere. For example, if VNGOs are not allowed to discuss the impact of a certain state-run mining site in Vietnam, they can attend oversea workshops and fora to discuss the project; they use the international sphere as a space to negotiate a local or national problem, rather than internationalizing an issue. The latter happens for transboundary topics, with the most prominent example being the Mekong River. Due to the international nature of Mekong water management problem, the solution needs to be cross boundaries, too. Here, the landscape becomes an actor and defines a network, namely, the Mekong.

In the matter of scales, all organizations work on the local, national and international scale but with different emphasis and in different capacities. All organizations receive at least some international funding and have networks beyond the Vietnamese border. All organizations have projects for policy advocacy on a national level, even if this is not the main approach of their work. And all organizations implement projects locally. The distribution across scales trait might also be due to my bias when selecting organizations to work with. Organizations that work exclusively on a local scale would have likely gone under my radar, as they are not connected to the national networks via which I got to know potential interview partners.

Due to the authoritarian nature of the Vietnamese government, cooperation with either local authorities or at least one ministry seems necessary for NGOs to operate, especially those that advocate for policy reform. NGOs usually cooperate with the ministry responsible for the topic, although informal networks are also formed with ministries that are in a key position in the Vietnamese ministerial hierarchy. On the local level, cooperation is also important since local authorities decide on the scope of work in their territory. The quality of the relationship determines if visits to project sites are allowed, if problems could be openly discussed, and if meetings could be held. When working with the government directly, the central committee is perceived as a more progressive actor than the National Assembly. NGOs actively select allies within the government system. They understand the heterogeneity of what look to be a homogenous block and learn to navigate the system through their networks.

NGOs can be divided according to their role of and connections with activism on the one hand and businesses on the other. Opinions on activists run the gamut from being important partners with a similar mission to being important for social development (but there should not be any network involving both activists and NGOs); to a clear dismissal of activists' theory of change, framing them as counterproductive for positive change or ineffective with their actions. Activists and their networks are usually less institutionalized than NGOs, and they often form and dissolve their networks around a specific cause, rather than maintaining a long-term advocacy project. An activist network that has a more advanced form of institutionalization in Vietnam is Green Trees. Green Trees is not officially registered but still perceives itself as a non-profit, civil society organization. It was started in 2015 from the movement against felling trees in Hanoi, and its initial network was a Facebook group. Several individuals continued using this group to organize protests and write reports and petitions to government officials on matters of environmental concerns, mostly in and around Hanoi. Their members perceive themselves as activists; they are partly under surveillance and actively persecuted by Vietnamese security and police forces. Their narrative on environment does not differ from that of the mainstream government. They use the frameworks of sustainable development and green economy, including embracing the growth paradigm and natural resource framing. However, their practice of these narratives makes them a security threat in the eyes of the Vietnamese government. An informant revealed that they would like to work with the government, not necessarily to question the legitimacy of the ruling party. However, the way they build networks and implement activities outside of the permitted institutional lines, seem to make cooperation with the Vietnamese state impossible. They use the government's narrative and turn it against them by showing how the state is not acting according to its own words. Green Trees also introduces words such as "human rights" (*nhân quyền*) and democracy (*dân chủ nghĩa*) alongside the Sustainable Development narrative, thereby partly transforming itself to become a government critic.

The level of institutionalization that NGOs function is set by boundaries. In this respect, NGOs are not strong narrative makers or norm developers; they do not use the proposal, enforcement and implementation power to make new norms, but to interpret existing ones. NGOs are more focused in their role as service providers, informants and advocacy organizations. They are actively engaged in information politics by gathering data and using them for policy advocacy work. NGOs see policy advocacy as an important role and actively connect themselves to political decision makers. For example, the CCWG hosts an annual pre-meeting for the Conference of the Parties (COP) to inform the Vietnamese delegation about priorities of non-state actors. The institutionalized access they gain through continuous relationship and network building with international, national and local actors grants NGOs access to powerful institutions that others, such as communities, do not have access to. NGOs, therefore, are in a powerful position themselves and can easily perpetuate systems of oppressions. This is happening in Vietnam, too.

The CCWG is one of several official networks for NGOs in the environmental field; the others are, for example, the Vietnam Sustainable Energy Alliance, the REDD+ network and the Vietnam Rivers Network. NGOs that are organized in these networks include both INGOs and VNGOs. As official networks are part of the support strategy that INGOs implement for capacity building and advocacy work, narratives and language of the international community are less connected to the international networks when VNGOs function outside the sphere. On a national level, VUSTA has initiated networks, but these are perceived as "must-go-to" events rather than lively forums for exchange and cooperation due to VUSTA's bureaucratic nature (Ortmann 2017). The VUSTA network meetings are, nevertheless, used as a space for advocacy work, which Yasuda (2015, p. 17) observes as the main function for networks. For example, the CCWG hosts conferences and publishes information and policy recommendations on Vietnam's nationally determined contributions (NDCs) for climate summits jointly with all member organizations to increase their leverage and pressure on the government institutions involved.

The CCWG is also organized under the VUFO-NGO resource centre. The resource centre was established in 1993 with a paid secretariat to organize and represent INGO interests in relation to other development actors in Vietnam, mainly VUFO. In times when regulations become stricter, the VUFO-NGO resource centre has assisted INGOs in obtaining and interpreting new information and sharing practices on the new decrees.

Ortmann (2017, p. 52) claims that building networks to improve capacity building and as advocacy tools are successful. He even sees their success as a potential "threat to power holders". Hannah (2007, pp. 223–26) also sees an element of empowerment of network building as the networks help exchange information, technical knowledge and habitus, thereby supporting new or smaller NGOs. However, he also sees network building in a negative light. The officially established and even partly internationally funded networks can also act as a space of surveillance, competition and exclusion for NGOs that do not wish to follow certain norms. This is true for one of the case studies in Chapter 5.

Unofficial networks are more spontaneous and rely more on personal than on institutional connections. They cross scales and government and non-government boundaries, but they serve similar purposes as the official networks: access to information, knowledge and advocacy work. Little attention has been paid to these unofficial networks, but they surfaced in the interviews and in the participant observation when solving problems like building trust with government authorities and communities. In places where official networks seem to be not workable, unofficial ones fill in. NGOs become both brokers of development and brokers of resistance. While NGOs, such as the Centre for Sustainable Rural Development (SRD) and C&E, do not criticize the Vietnamese governance system are understood as brokers of development, others become brokers of resistance. In their connectedness to local communities and the national state, NGOs can translate local resistance (Scott 1985) into formalized disagreement vis-à-vis the state to advocate for policy change.

Generally, besides discussing the different kinds of institutional cooperation, it is important to underline the importance of personal and individual connections as a parallel network that enables and informs a lot of NGO work. This is true for those who have moved across positions in NGOs, or from the governmental to the non-governmental side, and when making decision on fund allocation. Corruption and social obligations to family and friends play a factor too. As mentioned earlier, case study 3 in Chapter 4 focuses on its personal networks instead of its institutional networks; other NGOs are less outspoken about their personal ties as they fear a breach of compliance with codes of conducts.

As the role of the market and private as well as state-owned businesses have grown in Vietnam since the 1980s, the economy as a whole and individual businesses have taken on an important role as actors in environmental narratives. The economic system with its central role and inherent assumptions of growth and deregulation, is a scene-setter that establishes path dependencies for discourses. Its close-knit relation with politics gives it power to influence narratives. Looking at individual businesses and the structure that they are functioning within, we can see the perpetuation of narratives and economic practices.

While not all NGOs cultivate the same networks, they nevertheless help us to understand the concept of NGOs in Vietnam per se. The official or semi-official spaces for exchange among NGOs make it clear which organizations they regard as allies: either because they share similar theories of change and approaches to work, or because they face the same battles with institutional challenges imposed on them by the state (which then again is connected to the state's understanding of NGOs). At the same time, the networks that NGOs formed show which characteristics they seek in other organizations and through this process illustrate what matters to them. Surely NGOs and non-NGOs are not the only concepts to understand non-state actors in Vietnam. The term civil society has a prominent position in state-society debates and has been widely discussed in scholarships on Vietnam. It is timely to investigate if civil society exists in Vietnam and how it is related to NGOs.

Civil Society—(How) Does It Exist in Vietnam?

Following the self-definition of actors, this section focuses on NGOs that perceive themselves as part of CS, CSOs or NGOs. For example, in its strategy paper, GreenID (2016, p. 8) defines itself as "the leading credible Vietnamese civil society actor promoting sustainable energy sector development". CCWG and the Center for Environment and Community Research (CECR) distinguishes between NGOs and CSOs: international organizations are considered as NGOs, while Vietnamese institutions are branded as CSOs. A clear differentiation of roles and responsibilities is not spelled out, which makes scales the only differentiating characteristic. For example, CECR (2015, p. 5) states, "Use the competencies of NGOs and CSOs in facilitation, capacity building and joint monitoring throughout EIA process".

The self-definition of actors as part of CS or CSOs is partly a product of the international discourse. Both terms—CSOs and NGOs—do not have a fixed translation in Vietnamese and even remain untranslated in many publications. NGOs often remain untranslated, but if there is a Vietnamese equivalent, the terms used are *tổ chức phi chính phủ* or *tổ chức vô chính phủ*. *Xã hội dân sự* and *xã hội công dân* come closest to a literal translation of civil society, and these translations are used in NGO publications and rarely in government reports. The latter prefers referring to the socialist-derived terms "social organizations" (*tổ chức xã hội*), "socio-professional organizations" (*tổ chức xã hội nghề nghiệp*) or "political-social organizations" (*tổ chức chính trị xã hội*), which also includes organizations outside of the state apparatus as far-reaching as mass organizations. *Tổ chức phi chính phủ* and *tổ chức vô chính phủ* have a connotation of anarchy in the Vietnamese context, and therefore are hardly used (Nguyen 2012, p. 25). The most commonly used term remains socio-political organization in Vietnamese and, to some extent, the English translation.

Bui (2013, p. 78) sees CSOs as distinct in their organizational structure as they are shaped by characteristics of volunteerism, autonomy and self that exist within a set of shared rules. They stand in contrast to NGOs, which show a high level of organization, paid work and less autonomy due to the need to be recognized as an entity by the state. But these questions remain: When is a CSO truly autonomous? How do its connectivity and networks look like? How proactive is it in taking action if a threat to livelihood forces marginalized groups to act? And are groups which turn violent, either in language or action, no longer CSOs?

Several scholars use the Gramscian framework to understand the concept of civil society. This notion is contested by other scholars, among them Escobar (1995), who sees CS as a critical component for inducing societal change. Neglecting any meaningful agency of marginalized groups renders them even more marginalized and overlooks the power of everyday resistance that was touched upon in the discussion of development (Scott 1985) and is not compatible with social realities in the Vietnamese context (Kerkvliet 2005). A definition of CS, therefore, does not rely on its relation to the ruling class but on its practice, which are characterized by acts of resistance and associations.

The definition of CS becomes less clear as the scale and space covered increase. None of the definitions above are wrong, but they do not capture the essence of the Vietnamese situation and the self-understanding of organizations in the context. Therefore, this chapter follows Hannah (2007)'s argument of analysing activities, roles and networks which are functional and concrete, but with one restriction. The restriction is the frame of the Vietnamese state (a late-socialist state, a market-based capitalist economy, history and presence of cross-border and cross-boundary flows) that is informed by political, social and cultural practices.

Within this context, Hannah (2007) and Wells-Dang (2011) consider Vietnamese CS as a network. Hannah remarks that defining CS always requires an understanding of its agenda and its political positionality. Civil society emerged as a tool by experts who saw their development goals getting out of reach. NGOs themselves use the CS concept to counter critique and prove that participation could be enhanced within the existing development framework. It seemed like a one-size-fits-all solution for donor agencies and those supporting the discourse to strengthen governance and the market economy. The term civil society, therefore, became interlinked with the development agenda (Hannah 2007, pp. 66–67). Hannah (2007, p. 6) continues to define CS "in terms of the activities and roles of various social actors, rather than in terms of specific forms of organizations, associations and institutions". Wells-Dang (2011, p. 3) follows Hannah's approach of conceptualizing CS as a form of network-building instead of as organizations characterized by autonomy. Seeing it as a network extends the analytical capacity to beyond organizations. CS as a network includes institutionalized actors, and Wells-Dang (2011, p. 289) argued that individuals are also crucial actors who are highly relevant in making networks independent of their institutional affiliation.

As a structural problem, collaborations and networks that are highly criticized will most likely not get registered nor receive permission to operate. Therefore, they will be excluded from not only organized networks on the national sphere, but also most funding or donor agencies which only work with registered organizations. First, they are required by their Vietnamese registration authority to account for their activities and money flow. Second, their funding guidelines from the head office require tax invoices. Consequently, non-registration leads to a disruption on the international scale, too. These groups can nevertheless function, but they are not associated with CS.

Beyond Hannah's and Wells-Dang's approach, other scholars like Joerg Wischermann and Oscar Salemink contest the meaning of CSOs. These two scholars are at opposing ends of the debate in embracing vs. rejecting the term CS, respectively. Their differences lie in naming organizations in Vietnam that are not working directly under government institutions, and whether to extend the definition of the Global North to include other forms of CSOs and NGOs, or if a new definition and term should be established for the Vietnamese context. While both scholars have similar descriptions of actors and their networks in Vietnam, they disagree on the framing and naming of these processes. Wischermann (2017, p. 350) uses a wide definition on the terms CS and CSOs, and his own coinage "civic space" and "civic organizations", to include organizations from business associations to NGOs. At the same time, he recognizes the limitations of this wide usage but, nevertheless, finds it appropriate for the Vietnamese context. He accepts the fluid boundaries of spheres and understands the definition of CS and CSOs as inseparable from the state in the Vietnamese context.

Salemink rejects the terms CSO and CS for the same reasons that Wischermann uses them. Salemink argues that the interlinkage between state and society in Vietnam is so strong and multifaceted that these terms with their narrative concepts adapted from Western democracies are not applicable in the Vietnamese context. He further argues that these terms are imported and that the actors involved are much more diverse and are overlooked by conceptual frameworks. Similarly, Wells-Dang (2011, p. 169) finds that Wischermann's approach excludes individual activists, dissidents and protesters, and although these actors play an important role in Vietnamese politics, they are not the focus of this research.

Because of the heterogeneity of CS, it is impossible to discuss a state's response or attitude towards them. Instead, reactions and the level of acceptance of CS activities and their incorporation into the state's governance framework depend on the specific actor's positioning towards the official narrative and practices. CS may be instrumentalized by the government, but they are not necessarily so if their ambition is to make systemic changes. The level of institutionalization reveals the knowledge, capital, networks and aligned norms of CS actors and the state.

As this research focuses only on organizations that are institutionalized, using the terms CS and CSOs is appropriate. However, if the whole actor-networks existing in Vietnam were to be analysed, the use of these terms would not be sufficient. When collecting data for this research, the current conceptualization of CS and CSOs proved insufficient to grasp the community-based organizations, initiatives, village networks and spontaneous actions. These organizations are not represented in this research as they do not receive research permits and not involved with NGOs; hence, the terms CS and CSOs are used in a limited capacity. An all-encompassing term for non-structured and institutionalized organizations—"civil society organizations" (CSOs)—is preferred.

The next section illustrates how NGOs perceive themselves, how they build their networks, and what makes them "non-governmental". It shows the need to re-centre the state-society relations and to look at the negotiation processes between actors to understand environmental action. At the same time, this case study looks beyond the Vietnamese nation-state to colonial entanglements and processes in the global context.

Case Study 1: Navigating Relations in the Mekong Delta

"If you want to understand climate change and environmental action in Vietnam, you have to go to the Mekong Delta", the director of a Vietnamese INGO told me when we met in the meeting room of her organization. As she was known for her leadership, expertise and authority, I immediately agreed and asked if I could visit the delta with her organization. A few weeks later, I found myself standing at a crossroad in Binh Thanh district in Ho Chi Minh City waiting for a minibus that would take me and a delegation of five other people to a project site in the delta. The trip would illustrate how state-society relations work on an institutionalized level, and how personal connections and positionality shape the relations. Official networks and unofficial networks as well as self-definitions of the NGOs and their external definitions would eventually illustrate the negotiation processes between the state and society.

The NGO in this vignette was founded by a philanthropist in Europe in the 1970s. When it was first started, it was set up as an organization purely for poverty relief in the Global South. Over the decades, it has changed its scope of work, although reducing poverty remains one of its missions today. Besides women's rights, land and climate, politics and economy and emergency relief, its mission has broadened to social justice. This NGO was the first INGO to move its headquarters to the Global South, and one of the first to appoint a Vietnamese person as the country director. Its funding relies largely on individual and institutional donors, the latter being mainly development aid agencies.

The country office in Vietnam has existed for several decades. In the overall organizational structure, this NGO has a global strategy as well as different guidelines for external projects and internal management. For example, the NGO promotes feminist leadership throughout the organization but allows room for the country offices to adapt and implement these principles to their own context. The organizational structure in Vietnam is complicated by the legal framework. Although the country office in Vietnam relies on its "sister" organization, it is a legally independent entity.

The province we were headed to that day is an established project site for the NGO. They have two relational officers in the province who oversee the implementation of the projects. The province is geographically far from Hanoi and the head office; it takes a two-and-a-half-hour flight and another approximately five hours by car to reach there. Even from Ho Chi Minh City, where the NGO has a small presence, the journey requires a ten-hour car ride. The local relational officers, therefore, play an important function in maintaining relationships with government institutions and communities and managing everyday matters. Both officers are from the communities where they work, and they are very knowledgeable about the local context and networks. They actively translate local circumstances and needs to Hanoi and help to explain the NGO language and requirements to the communities.

I spoke to members of the communities about three projects that were initiated there. The first project dealt with the issue of plastic waste. It aimed to raise awareness of waste management through education. The second project transformed forest management practices by empowering forest dwellers to better manage the forest sustainably. For this project, computers were installed in the municipal buildings, and staff were trained on how to use computer programmes to manage forest data. Livelihood programmes were also further developed to reduce poverty among the participants. The third project was on safety issues in the communities. While all three projects used a social approach towards environmental topics, the third project reveals the most about human nature and power relations. The province has been characterized as an area where sustainable development and social development could be applied against poverty.

A delegation of three staff from Hanoi and Ho Chi Minh City, three invited external experts and I arrived at the provincial party school. It was early morning but hot; the ceiling fans were turned on at their maximum capacity, but this did not seem to be enough. Everyone was busy getting the location ready for the training that was about to start. I looked around the place. It looked like a typical government building in the countryside, not much different from those in the Red River Delta and Mekong Delta. The portrait of Uncle Hồ (Hồ Chí Minh), with a yellow star, hammer and sickle above his head, hanged high and oversaw the room. Party slogans framed the stage and the backdrop read: "training". The wooden chairs were as always uncomfortable. We were clearly in a space of government authority.

The participants arrived, and they seemed shy or even intimidated: some spoke to each other but most of them sat down quietly. All of them were in a position of power: some were local party leaders; others were delegates from the local chapter of the women's union, the Hồ Chí Minh Youth League, or the Farmer's Union; or they worked for the government administration at the local or provincial level. They were able to improve local livelihoods and were key persons in understanding the realities in the communities. All the participants could be considered as local because they came from or resided in a province away from the centres of political and economic decision-making. At the same time, they were connected to the system of power and therefore differed from other "locals" who lacked direct access to decision-making institutions. The multifaceted identities of the participants show the complexities of power relations across scales and the diffusion of what are state and non-state actors (Schwenkel 2020, p. 239).

The location of the training and the participants involved did not seem to influence how the NGO staff carried out their role. Previously, I have observed the delegation interacting with the local community of farmers, women and the disabled in North Vietnam and there seemed to be no difference in their interaction here. One of them, whom I called Huy, had an air of superiority. He was constantly interrupting people and directly dismissing their ideas by bluntly stating "*Không phải*" ("Not right"). He raised his voice and ridiculed other people's contribution. In contrast to Huy, Giang (all the names in the NGO stories have been changed) showed a very different way of interacting with other people. Rather than stating her opinions, she asked questions. Instead of interrupting others, she encouraged them to elaborate on the points they made. She greeted everyone in the room and asked how they and their families had been. As she had been to project sites previously, it was obvious from the smiles and affectionate handshakes that the people were happy to see Giang again. Giang was aware of Huy's behaviour and intervened whenever she felt his response was inappropriate. She relativized his harsh criticism, laughed to ease tensions and at times even laid her hand on his shoulder to calm him down. Giang and Huy and their interaction with each other showed how personal characteristics and self-perception of one's position can result in two very different approaches to working with local communities within one organization.

This difference became apparent when I spoke to both Giang and Huy. Giang told me how much she loved the countryside, being a rural girl herself. Although she moved to Hanoi in her younger days to study and now lives in one of the modern high-rises, she still owns a house and a garden in her husband's hometown. Her husband does not belong to the Kinh majority, but he is of ethnic descent. According to Giang, this relationship helped her to better understand the livelihoods of ethnic minorities and value their knowledge of nature. She said her favourite part of her job is to visit the communities and work with them. She can translate her Hanoian INGO identity and language into the local context. Her colleagues turned to her whenever there are problems with the local communities. During our interaction, she also explained contexts and translated for me when the local terms and languages did not match with my Hanoian vocabulary. Giang reflected on her own position and what it meant in relation to the person she was interacting with. On the other hand, Huy was rather reluctant to talk to me. In his comments, it became clear that he thought other people knew less than him about certain issues, and he openly wondered about women discrimination when we discussed patriarchy in the Vietnamese context. The rest of the team gendered those behaviours with jokes and comments. What we can take away from these different behaviours is that self-identification in NGO work matters. How the staff perceive their role and behave towards other people and institutions shape the NGOs' positioning in their networks.

The whole delegation was there to train the local government's key persons on social auditing. It was a session to learn about the methods of social sciences used to better understand the needs of the communities. One expert who was invited to conduct the training session was Minh, a lecturer at a university in the North, who specialized in forestry and the environment. However, he was invited not because of his specialized knowledge, but due to his knowledge on the methodology of social sciences. Minh conducted his training sessions like a university lecturer. He gave a frontal lecture on the first day, occasionally asking questions to the audience and answering them in a rhetorical way. He established his authority as a teacher, although his audience were not undergrad students but government officials.

Minh was openly critical about the executive part of Vietnamese government in his lecture. He started by theorizing about the government's role in a society. He referred to Marx but also related it to the Vietnamese reality. Minh used the metaphor of a boat: The state is the boat's captain, but it is the people on the boat who decide where to go. The government has the responsibility to "understand each other, share information, sovereignty, seek understanding, analyse". If the state asks the people to "*chịu*" ("endure"), then it means that the state has not done enough or the right thing for its people. Referring specifically to taxes, Minh stated that taxes were necessary for building social structures, but a trustworthy tax system would also require transparency, and corruption is a big problem everywhere in Vietnam. If the government is interested in its people, it needs to welcome criticism and not dismiss them. At this moment, Giang seemed to notice that the room had become tense, and the atmosphere had turned uncomfortable. Everyone was silent. She briefly interrupted Minh with a smile and said: "*Nghiên cứu viên thì nói thật*" ("This researcher speaks the truth"). It seems that by supporting his argument and at the same time calling him "the researcher", she abstracted his ideas to an academic level with authority, not addressing him by his name but rather his position.

Minh continued his lecture and argued for development. He said the more developed the country becomes, the less corrupted it gets. He asked the audience for their opinions. Nobody responded and those whom Minh directly addressed seemed terrified and speechless. Minh became impatient and instead of waiting for participants to speak, he continued his one-sided lecture. "The government also makes mistakes", he said. "That is why it is important to keep a good relation to the people. Only because of their support, the state is able to exist. To contribute to development, the government needs to understand and to understand requires data and statistics. That is why this training introduces auditing processes." The framing of this project is modernist and rooted in an understanding of science and technology.

Minh further referred to socioeconomic development as the main principle, and sustainable development is to be seen within the wider socioeconomic development framework. He thereby replicated the government's narratives and directives. He called for more participation from the people, yet at the same time said that citizens *"không biết đủ"*—they do not know enough to understand the overall complexity, which is the reason why experts are needed in policymaking. Citizens still follow the approach of "Only know it after you saw it, only see it after you knew it". Experts, in contrast, would be able to balance the subjectivity of diverse opinions. These factors are to be kept in mind when conducting a social audit. Minh, thereby, rhetorically strengthened a hierarchy of knowledge that is preeminent in the Western academic system. Subjective vs. objective knowledge, laypersons vs. experts, experience vs. data are dichotomies that serve racist, gender and socioeconomic stereotypes exported by colonizers to other parts of the world (Kilomba 2009). Laypersons vs. experts is, however, a distinction made by not only the Western academic system, but also the Chinese mandarin system. Which of the two has a stronger influence on Minh's thought cannot be determined, but either way, this narrative is uncritically reproduced here, although the lecture itself is critical towards national authorities.

In the same way, Minh did not apply his analysis of the national government's power to other international institutions. Referring to the World Bank, he explained, "When the financer wants to invest, then there needs to be change" and "if you want to play with the world, then you have to play according to rules". He saw "international rules" as legitimate and something that Vietnam needs to comply with in order to "modernize" (*hiện đại hóa*). In the same stance, Minh depicted "the West" (*Phương Tây*) as a morally superior role model in the fight against corruption, which is widespread in Vietnam. Studies (Taenzler, Maras, and Giannakopoulos 2012; Transparency International 2021) show how this claim does not hold up, but still, the narrative has been internalized in Vietnam.

Another stereotype about a morally superior Germany came up during a discussion about men's violence against women (MVAW). When discussing the topic of what a safe community is and means, MVAW is mentioned as an issue. The discussion is gender-sensitive but still reproduces gender roles and those of sociocultural spaces. Minh pointed to me and mentioned that in the country where I came from, there is no MVAW issues. I disagreed and tried to explain how MVAW is still a widespread problem in Germany and Europe. He had no reaction to my statement; instead, he continued discussing the situation in Vietnam and tried to convey his message of MVAW as a problem through jokes, leading me to question his understanding of and sensitivity to the topic, especially with a gender-diverse group of participants.

One participant later asked if we had to include the soccer field when analysing safe places for women and children because most girls do not play soccer. Giang intervened as she played soccer all her life and was an exceptional example. Connecting women and children into one group for analysis targets the problem of marginalization by a male-dominated society. This does not, however, question power structure by acknowledging women's agencies and the need for empowerment. Instead, the discussion is led in a way that sees women in a position of lesser power, and that those in a more powerful position should treat women better but without changing the power structure per se. This is surprising as the NGO has an openly feminist agenda. But the translation of these values into work on the local level is not taking place in this project.

Minh talked about the importance of participation in his lecture. Although he mentioned "*tổ chức xã hội dân sự*" (civil society organizations) and NGOs as important factors for a participative society, his lecture is not participatory. He talked for long stretches of time and whenever he asked questions, the others from the delegation and not the participants replied. Hence, narratives are not translated into actions.

Although Minh was an environmental expert, he did not make any references to safe communities, development and environment. The participants hardly brought up environmental concerns and did not seem to connect safety with environmental crises. Besides hygiene as a topic that implicitly connects environmental topics to health and female hygiene, there is no proactive narrative of the socio-ecological connection. Sustainable development is included under the paradigm of socioeconomic development without an explicit environmental reference.

When I spoke to the participants bilaterally, they mentioned that they lacked an understanding of what "sustainable development" means. Although red banners on the streets of the communes proclaimed "sustainable development" for the province, yet none of the people I talked to can explain what this term means. This term is mentioned by the government as a framework for long-term stability. One respondent thought that it means if you have ten million Viet Nam Dong (VND) in one year, you will have more than this amount the next year. Another one saw it as a long-term plan for development and poverty reduction. This term seems to be closely associated with temporality. Despite not knowing what sustainable development means, all the respondents knew that it affects their future. Although the people I talked to think that sustainable development is supposed to solve problems that are related to their agencies, they did not see their agencies as part of the plan. They perceived sustainable development as positive and life-improving.

That does not mean, however, that the people I talk to did not realize environmental problems. They perceived weather changes and the need for action. All the people I interviewed pointed out that the temperature has risen remarkably on average, especially during summer. Alongside the temperature rise, everyone reported an increase in drought frequency due to less rainfall. Consequently, water canals have dried up and farmers witnessed saltwater intrusion. This destroyed pipes and canals, which are often clogged as not enough water flows through them. Fish and shrimp also die more easily. Farmers also reported that their livestock such as pigs and chickens got sick more often due to temperature rise, and they require more medicine, and this affected human health.

Regarding environmental issues, the respondents preferred to discuss "environmental protection" (*bảo vệ môi trường*). Similar to the interviews with the NGO staff, the respondents perceived the environment as the space of interaction between humans and nature. Environmental protection means changing human actions to reduce negative effects on the environment. This includes proper waste management and wastewater treatment in both households and businesses. Environment is also often connected to cleanliness: the "*xanh sạch đẹp*" ("green, clean, beautiful") slogan, which is widely used by the government, seems to have influenced the people's mindsets. This slogan is seen on propaganda posters to encourage citizens to protect the environment. It is also seen on signboards at green spaces to remind the people to keep the area green, clean and beautiful. It gives the notion that the environment is tamed, orderly and has aesthetic value (Harms 2016).

The government should take the lead in environmental action with the citizens following suit since the responsibility lies in higher levels of the hierarchy. Although the respondents recognized the importance of their daily actions on the environment, they did not regard themselves as the main actors in solving environmental crises. They recognized the responsibility of the state to maintain a clean, green and beautiful environment for the interest of the public.

Despite these common observations, the people responded differently when asked how the environment has changed. Most understood this question to refer to not only climate change, but also waste management. Some of them praised how much cleaner their commune has become over the last few years. The people are more aware of recycling now, and this creates a cleaner living space. This development is brought about by the *Nông Thôn Mới* Program that was initiated by the Vietnamese government in 2011. Its goal is to "effectively build sustainable rural areas" with the construction of "model residential areas and gardens; carry out the One Commune One Product (OCOP) Program; apply science and technology to new countryside construction, and associate rural tourism with new countryside construction". This programme frames sustainability in an understanding of modernity and technology with a focus on effectiveness. It provides infrastructure for transport, education, health to half of Vietnamese communes as of 2020. Its slogan "*sáng, xanh, sạch, đẹp*" emphasizes a positive outlook.

Other communes, where the *Nông Thôn Mới* Program has not been implemented, expressed their frustration about waste being discharged haphazardly into the environment. Although they praised the efforts undertaken by the government, they felt that these were insufficient. The government was aware of many problems, but they did not follow the advice from researchers. One person mentioned that this is due to the corrupt system where the rich have no interest to help the poor. The respondents seemed to view environmental and social factors as closely related to the *Nông Thôn Mới* Program.

The training continued as the participants developed a plan and social audit system and visited selected communes to develop policy recommendations to make the communities safer. We visited different communes, among them ethnic Khmer villages, to evaluate the people's opinions on safe communities and spent the next two days gathering data. Some language barriers existed but were overcome with the help of key persons from the village. We approached the Kinh communities with the same concerns as the Khmer, but with additional questions on their access to Kinh institutions and structures. In each commune, three focus group discussions took place: one with the commune administration, one with women and one with men. This approach is supposed to encourage the interviewees to share their thoughts more freely.

Two local project officers from the NGO coordinated the work in the communities. They served as network brokers by introducing NGO staff to the community, choosing the best place to hold the focus group discussions and arranging meetings with key persons from the community. They also arranged meals for both interviewers and interviewees to eat local produce together. This helps to bond us because respondents who shared a meal with us seemed more willing to share their opinions. The NGO worked with the communities by gathering information about them and using their position of power to handle the local villages and towns. However, this practice is implemented differently from person to person.

Relations with cooperation partners are also reflected in the relation between the NGO and the local and provincial authorities. An unofficial part of the training programme involved a dinner and karaoke session with the district government, which is responsible for granting permission for activities to the NGO. Although this NGO is known for being critical of the government, it is able to build trust with key government institutions and the people on the provincial and communal levels. It recognizes the existing power structures and seeks to change them by cooperating with the government. It aligns with government narratives just like China's "green consensus" (Arantes 2023). Like Vietnam, the Communist Party in China has stated the norm for its environmental rule, under which only decentralization of tasks can happen. Critique is directly voiced within the cooperation with the state and relies on the existing individuals from the state and NGOs. The state defines what the NGOs can do and what networks they can establish. The NGOs must adjust their work according to the conduct allowed, but within that space they can navigate the contents of their projects.

The last day of training was dedicated to working with the state and locating the NGO as agent within the system. For the closing ceremony, political decision-makers were invited to the presentation of the social audit results and to understand how people in their constituency perceived safety. The training participants presented the results and suggested a few policy recommendations. One observation was that communities that participated in the *Nông Thôn Mới* Program reported less issues with safety. Another was that administration officers and those from the communities showed differences in their perceptions about safety, which underlined the need for exchanges to understand the people's needs. Despite the NGO's efforts for cooperation, there is still limited interest from the government to take up the recommendations. This might be due to limited capacities and unclear responsibilities. But it also relates to hierarchies and the general state-society relationship: although the NGO is perceived and perceives itself as external of the state, it is dependent on it. The NGO staff are considered as outsiders who are granted cooperation by individuals in the system who favour network building.

Conclusion: Re-centring the Hyphen between Human-Nature and State-Society

This chapter explains why this book focuses on NGOs and, to a much lesser extent, CSOs. NGOs within the Vietnamese political-economic context are defined by their self-identification, their structural traits and external regulations, and their networks. Although NGOs are similar in their positioning, they are a heterogenous group. They have diverse functions: narrative makers, norm developers, service providers, advocacy organizations, translators, representatives, intermediaries, gatekeepers, informants, sociopolitical makers and discursive actors. The different negotiation processes of state-society relations also contribute to the heterogeneity of the NGOs.

Aside from the structural commonalities, the ways in which organizations formed their relations vary: personal relations matter as much as institutional ones, individuals who are both in the state and society realm and their overlapping roles and positions matter, and the processes across scales from local to the global need to be understood in the Vietnamese context and its environmental rule. The case study has demonstrated the realities of the project on the ground.

Since 2021, the state surveillance of NGO activities has increased dramatically. It remains to be seen how and if NGOs can adjust their self-positioning under stricter state control. INGOs need to ask themselves the same question: can they realistically comply with the values and policies that the headquarters demanded from them? And how has the risk assessment of connecting scales changed? What we can see now is a further distancing from activists and an economization of the NGO sphere. To avoid state regulations, the registration of social enterprises or business entities is seen as a way to navigate power relations. Consequently, NGOs might even move further away from scholarly concepts, recentring the need for alternative conceptual frameworks.

The vignette has also given us a glance into the importance of following processes across scale and including power relations in the analysis. While this chapter has focused on establishing the actors for the research, the next chapter connects them to the narratives more closely. With this knowledge of *who* is establishing and carrying out environmental policies, we can now turn our attention to *what* is said and understand *why*.

4

Moving beyond the National: Positioning Actors and Narratives across Scales

Chapters 2 and 3 focused on Vietnam's national context, and the state governance and actors as well as society are mainly involved in national politics. But to understand the connection between narratives and power, we need to move our analysis beyond the national context. The nation-state is indeed very important in environmental governance—in previous and later descriptions we can see how the institutions form actual realities on the ground. Nevertheless, Vietnam as a nation has always been embedded in transnational flows, connected with international actors and formed part of processes that cannot be assigned to only one scale.

As Tsing (2005, p. 1) states, "Global connections are everywhere" and contends that we reflect on our methods to find and understand the universal. We must resist the temptation to homogenize the global and heterogenize the local; instead, we must see "the global" as something that is co-made in the local, and vice versa. Sassen (2008, 2014) reassesses and disaggregates scales in her analyses because the presumption of them can bring biases into investigations. As an alternative, she scrutinizes foundational components as the starting point of her processes across scales; she puts theory at the end, and not at the beginning, of her work to truly understand the reason behind.

To avoid categorizations, I examine environmental narratives and study how they are coproduced by actors across different scales in Vietnam and beyond. My research shows that using some scales and categories is inevitable to understanding power relations. But we need to pay equal importance to processes that move across scales and to connections between actors in different geographic location. Therefore, this chapter expands the analysis on environmental narratives and the constitution of actors beyond the nation-state to the international level.

Grounding Contemporary Environmental Narratives in Historic Context

The narratives that exist today need to be understood in their historic context. Trade, migration, wars and colonialism have shaped narratives and our understanding of nature in Vietnam today. Buddhism, Confucianism, Daoism and many other belief systems came to Vietnam and formed unique philosophies in combination with the local animistic beliefs. These beliefs have demonstrated the nature-human relations. For example, the Yin and Yang (dương âm) belief in reference to Daoism is itself not part of a single belief system or religion, but rather the result of the different influences in Vietnam (Jamieson 1995, p. 16). Jamieson calls the balancing forces "one of the oldest and most fundamental elements of Sinitic influence" that informed the entire Việt system, governing everything from beliefs to household organization, and even the state. Every form of organization became a symbol of Yin and Yang, which formed the basic structure of the universe.

Another approach to understanding historic narratives derives from etymology. Weller (2006) looks at the understanding of nature from a Chinese linguistic approach, which is mainly influenced by scholars of Confucianist thoughts. Việt culture has adopted many Han-Chinese traits, including the linguistic term tian, or thiên in modern Vietnamese. While tian is translated as nature, historically it means heaven, and therefore shows differences in worldviews and meanings of the term. Weller (2006, p. 21) states, "Tian comes closest to the meaning of nature as an inherent force directing the world. Etymologically, the word initially referred to the sky, but it quickly took on broader connotations of heavenly power."

This holds true for Vietnamese language as well; *thiên* is literally translated to heaven or sky, but it refers to nature when combined with *nhiên* to become *thiên nhiên*. *Nhiên* means "course". The heavenly course, thereby, becomes nature. It covers the English understanding of nature, which is the primordial, unaltered and unchanged way of things. The broader use of the *tian* concept can be found in Chinese texts; for example, *tian di ren*, which means "heaven, earth and mankind", expressed a view on how the world is organized. *Tiandi* refers to everything but humanity (Weller 2006). This meaning is close to or at least compatible with environmental narratives in Europe and the dichotomies around nature and culture. Droz et al. (2022) explain that *ziran* is the term that nowadays encapsulates the English term "nature", and since the Sino-Japanese war, it has gradually shifted its meaning to the Western connotation that excludes humans, although in pre-Buddhist and Buddhist periods it referred to humans, too.

These two approaches exemplify how flows and international actors have always influenced the thought system and narratives in Vietnam. However, questions remain: how far has this thinking been translated into practice and to what extent does it continue today? To answer these questions, we need to search the archives, but the written sources have the disadvantage of representing the powerful and literate. It is easier, however, to move forward in time.

Beyond these early influences, colonialism and the French rule had important influences on the concepts of nature in Vietnam (Said 1978). The French influence was exerted mainly through colonial policies. The French colonizers replied on both the local and national elites as they did not intend to change societal structures beyond what was required for administrating and managing colonial policies. Still, the combination of direct and indirect rule was essential to the French power network because of the colonial administration's lack of funds and the need for people who understand the local language and culture. The elites, thereby, formed transnational networks of power.

The French's primary vision was for Vietnam to provide exports to France, improve Vietnam's transport infrastructure and focus on its plantation agriculture. The development of the agricultural economy had a lasting impact as this was continued even after the end of colonialism. Aso (2012) explains how this sector developed. The Farm School of Bên Cát was established in 1917 in Cochinchina to educate people on managing the Indochinese agricultural economy and farming techniques (Aso 2012, pp. 31–32). Its goal was to enable Việt people to develop the agriculture sector and attract graduates to work in plantations. However, interest in this school was rather low and most students came from families who owned small local properties. To these students, this Western scientific education enabled them to attain a certain social standing, analogous to the modern version of Confucian training. Even in the resistance against the French, modern scientific knowledge was seen as more useful than traditional subjects. While the French saw education as a means to increase profit from a well-oiled colonial extraction industry, Việt saw it as a way to independence and strengthening themselves as a nation (Aso 2012, p. 41). To the Việt, education was a means to understanding one's oppression while trying to gain independence by adapting ideas and practices from Marx to their context.

The administration managed the colony from Paris. There was some degree of autonomy due to the distance between France and Vietnam, and each governor left their own political mark during their time in Vietnam. Transfers from the colony to the power centre included not only resources, but also knowledge. French scientists and travellers were interested in local knowledge and transported them across borders. The Việt elites made observations and conducted research to rule more efficiently, rather than to understand the people's needs. These included not only specific matters of interest (e.g., botany and agriculture), but also general data on the residents, land rights, etc. Research on fauna and flora in the colony were conducted for scientific interest and then used to develop strategies for profit. This information, together with the personal accounts of the French administrators (e.g. letters and diaries in annual reports and correspondences between the colonial outposts), illustrates the perception of the environment in the colony, which was then reflected in the colonial policies regarding agriculture and export economy. Strategically, the Red River Delta was deemed as a more important place than Cambodia regarding relations to China. The cadastres, maps and other data-based material were used to build a strong state based on the written accounts that the mandarins had left behind on governance (Biggs 2010, p. 81). The archives of Indochina states, "The mountainous region which surrounds the delta does not lend itself either by its physical characteristics which make almost everywhere very difficult to access, nor by the conditions in which its inhabitants live, to the application of the same regime as the delta"[1] (Devos, Nicot, and Schillinger 1990, p. 16). The state was then adopted according to the different landscapes.

The French colonialists intervened significantly with Vietnamese landscapes. Morris (2011, p. 155) reminds us that landscape change had happened before the arrival of the French—but not at this speed and to this extent. The French introduced changes that differed from those previously due to the scale and the different narratives regarding landscape transformation. The French's ideology of exporting goods was based on the understanding of nature as embedded in colonial domination, enlightenment and capitalism (Behrens 2014). This results in high resource exploitation, enhanced marketization of Vietnamese agriculture, new clusters in networks of power, and a simultaneous transformation of social and environmental order.

Documents, such as diaries and letters from French bureaucrats and colonialists in Vietnam, showed the colonial rulers' limited perception of the Vietnamese environment. These sources described the economic and extractive potential that the colonizers saw in exploiting nature, discussing how to make use of natural resources to yield the highest gains. In addition, references to the environment as a space for war were made. Furthermore, a few diaries described how French personnel attempted gardening, including their stories of success and failure. Overall, the accounts and descriptions I have encountered are all related to making use of and extracting value from nature.

In the agricultural economy, the emergence of plantations was a major change in the characteristic of landscapes. The southern region of Vietnam was more affected than the northern region as rice production remained in the hands of the Việt. Rubber and its processed form, latex, were sought after for industrial production in France (Goscha 2016, p. 164). Other common plantation crops were coffee and tobacco (Freud 2014, p. 99; Boomgaard 2007, p. 178; Roubequain 1939, p. 109). The land used for plantation tripled from the beginning to the end of French Indochina (Jamieson 1995, p. 90). The introduction of plantations seemed random throughout the whole period of French Indochina. Aso (2012, p. 20) argues, however, that despite irregularities and the refusal of local farmers to expand their plantations,

---

1. *"La région montagneuse qui entoure le delta ne se prête ni par ses charactèrs physiques qui rendent preseques partout très difficilement accessible, ni par les conditions dans lesquelles vivent ses habitants, à l'application du même regime que le delta."*

(rubber) plantations in Cochinchina were part of the French project to create "indigenous capitalists".

Changes in land use resulted in a shift in land ownership and resettlement. The landowner derived marginal revenue from agriculture. The colonial administration transferred landownership to landlords who then rent the land to tenants (Luttrell 2001, p. 65). This centralization of landownership was supposed to change farming techniques and make crop selection easier for export businesses.

In the uplands, this meant that shifting cultivators had to turn to rice production (Boomgaard 2007, p. 223; Aso 2012, p. 20). The colonial administration and their scientists did not understand forest agrosystem in the mountainous areas (Thomas 2009, p. 127). Even today, the Kinh continue to lack understanding of agroecology, and they still blame ethnic minorities for deforestation, which pose major threats for the forests in Vietnam. The French exploited the forest lands, and they did not establish conservation measures and were not interested in communal forest management practices (Thomas 2009, p. 131). While the French acknowledged the harmful effects of deforestation, the responsibility for conservation measures was placed on indigenous forest management practices. As a result, the French kept using European techniques on the local forest, and this further worsened the problem (Thomas 2009, p. 108). The French jury, which oversaw agricultural products and industrialization in Cochinchina, declared:

> We know what these half-nomadic populations are, the most miserable in Cochinchina, who live two or three years on artificial fertility, then move to go further into the forest, marking every step with such unnecessary destruction to the cultures they practice that is fatal to their well-understood interests, and thus move away more and more from civilization and wealth, leaving between them and those an immense desert sterilized by fire[2] (cited in Thomas 2009, p. 108).

The integration of the upland society into the value chain aimed to market the colony and meet the demand from France and the elites in French Indochina.

---

2 *"On sait ce que sont ces populations à demi-nomades, les plus misérables de la Cochinchine, qui vivent deux ou trois ans sur une fertilité factice, puis se déplacent pour aller chercher plus avant dans la forêt, marquant chaque pas par une destruction aussi inutile aux cultures qu'ils pratiquent que funeste à leurs intérêts bien compris, et s'éloignent ainsi de plus en plus de la civilisation et de la richesse, en laissant entre elles et eux un immense désert que le feu a stérilisé."*

Measures were made to remove the uplands' attributes of remoteness and "backwardness" (Micheaud 2015, p. 344).

Additionally, the French colonizers perceived the farmers of the Red River Delta as more efficient—a perception that continues to live on among the Vietnamese today (Biggs 2010, p. 109; Roubequain 1939, p. 62). A resettlement plan was, therefore, implemented and peasants from the North were allocated to the South, leading to changes in social structure and conflicts, as well as merging in cultural understandings.

The Mekong Delta experienced the biggest changes to its canals during colonial rule because of the system of landownership, the expansion of plantations and the resettlement of Northerners to the South. But infrastructure programmes and the drive for urbanization also left their marks. Although these had already been under way before the French arrived, colonial narratives and business practices and modernization increased the speed of change and the method of implementation (Biggs 2010, p. 57). The landscape was changed rapidly through building canals and dykes (Duong, Safford, and Maltby 2001, p. 193; Roubequain 1939, p. 63). Two million hectares of floodplains were rebuilt into canal systems (Biggs 2010, p. 109). The French engineering plan differed to that of the Vietnamese in that the French measures included static waterways that required the enclosure of water flows. Biggs (2010, p. 43) also notes the importance of French technology: "Dredges chewed up fields, forests, and huts in their paths. More than gunboats, locomotives, or machine guns, the arrival of the dredge meant immediate ecological and social change."

As a result, we see different environmental narratives in current politics that place seemingly contradictory storylines next to each other. Contradictions between the Daoist view of seeing everything as interconnected and the separation of human from nature is visible in environmental action today. Although governmental reports refer to the economization of natural resources and the human responsibility to take care of nature through conservation measures, they also mention intrinsic values of nature in a few instances. For example, the National Strategy Against Climate Change (2011) states: "Development must observe laws of the nature, harmonize with the nature and befriend with the environment; economic development should be suitable to ecological features of a certain region, produce a minimum of waste, especially carbon, and strive for a green economy". In civil society, the protesters against the felling of 6,700 trees in Hanoi in 2015 combined the two narrative traditions. When the government planned to cut 6,700 trees in the streets of Hanoi to make space for infrastructure development, the people organized themselves on social media and on the streets. Their arguments included concerns about ghosts hiding in the trees and fear of revenge; specifically, they feared that human lives would be lost because the homes of the ghosts (the trees) were destroyed, bringing catastrophe to the whole area where the trees were located. At the same time, the people argued that they would lose clean air and the aesthetics of the city would be ruined. All these arguments, though coming from different understandings of the human-nature relation, were combined to counteract environmental destruction and influence Vietnamese environmental action today.

Persistence and Limitations of Colonial Entanglements

Power relations across borders need to be understood historically in the colonial context. Many (neo)colonial structures and institutions have left their traces in contemporary politics. Yet, few analyses or implementations of activities in Vietnam's environmental sector pay attention to this continuation. Again, the environmental rule approach illustrates the historical legacies and put them in the centre of analysis. I began the analysis by looking at donor agencies based in the Global North, how they create new dependencies and perpetuate power relations that are eco-imperialism (Gonzalez 2000). This means that countries and actors of the Global North make use of environmental politics to interfere with the sovereignty of countries in the Global South, a parallel argument to environmental rule within the Vietnamese context.

Colonial structures are, however, only part of international influences that shape environmental action. International actors enrich environmental narratives and practices with their financial resources and knowledge. Non-governmental organizations (NGOs) and civil society organizations (CSOs) in their policymaking processes create space for participation within the authoritarian context (Ortmann 2020). Practically, this means including and going beyond the West and the Rest argument that creates a dichotomy between hegemonic countries in the Western hemisphere (e.g., Europe, North America) and formerly colonized countries ("the Rest", as ironic commentary to the West's claim to power over them).

How the international actors frame their financial spending on environmental activities and development aid is crucial. Former colonial powers frame them as aid. Ayers and Abeysinghe (2013, p. 492) criticize this "aid" to former colonized countries as nothing benevolent. In fact, they argue that this aid should be considered as compensation for the countries who have hardly contributed to climate change and biodiversity loss, yet they face the most adverse effects from climate change. By framing the support as aid and deciding how and on what financial means it can be spent, colonial power reactivate their power relations, and this provides a legitimate reason for them to interfere in other countries' jurisdiction. This exterior decision-making power also influences the narrative frame of environmental projects. Furthermore, international actors contribute to the narrative-practice gap in NGO work. Vietnamese NGOs (VNGOs) either follow the narratives to be eligible for funding, or they avoid openly challenge power relations and quietly do things their own way. They navigate power relations by adjusting to the narratives for financial resources and networks.

Due to its colonial history and its struggle for independence, Vietnam is aware of these problems. To curb international influence, it officially "maintains a strict policy of 'protecting domestic politics' (*bảo vệ chính trị nội bộ*) from outside interference, seeking to closely control how foreign recommendations are implemented" (Weger 2019, p. 187). Vietnam is not a passive receiver of international influence; in fact, it has its own agency in choosing with whom it forms networks. Since *Đổi Mới*, Vietnam has been strongly committed to multilateralism and integration into the world. It is a member of ASEAN and the WTO, and it has increasingly sought a central position in international relations by hosting the Trump-Kim Hanoi Summit in 2019 and by becoming a non-permanent member of the UN Security Council. This is also reflected in its environmental policy. Official development assistance (ODA) became an important resource for the country in the 1990s, especially after the US embargo was lifted in 1993. Gradually, ODA changed its role from realizing infrastructure projects to focusing on institutional reforms (UNDP 1999, p. 26). Vietnam has also taken part in major UN processes since the Earth Summit in 1992, and it is a party to the Kyoto Protocol, the Paris Agreement, the Convention on Biological Diversity and many others. The processes leading up to these summits have been a point of engagement for several international organizations, including international NGOs (INGOs) and subsequently VNGOs, in influencing the government's policies, mainly through institutionalized networks. Collaboration in the international space has also created opportunities for advocacy work and critical engagement between state actors and non-state actors on the national scale.

International agreements that Vietnam has signed and ratified have an influence on its national policymaking. The government papers incorporate international regulations, and other actors like NGOs see international agreements as a framework to conduct activities. Schreurs and Economy (1997, p. 2) note that the influence of international politics has increased in the environmental field, and it continues to do so today: "[t]his internationalization of environmental politics is transforming the relationship among actors within and among states ... Agenda setting, policy formulation, and implementation are becoming increasingly internationalized". This increases the "dense vertical linkages between domestic environmental politics, state institutions, and the field of global environmental governance, thus [...] reshaping established patterns of policymaking within the state and beyond" (Falkner 2013, p. 257). Governments formed part of the international process of co-shaping and influencing domestic policy.

Most recently, a Just Energy Transition Partnership between the Vietnamese government, G7 countries and the EU had an influence on Vietnamese domestic politics. The partnership, which was announced in December 2022, aims to replace fossil fuel with renewable energy through a financial aid of 15.5 billion in loans, grants and investment. The agreement is currently in its very early stages of implementation. It remains to be seen how these large-scale environmental development projects materialize.

Donor organizations, development agencies and INGOs in Vietnam support the implementation of international frameworks. Donor organizations and development agencies differ from INGOs in that they are linked to government institutions, but they are often perceived as similar institutions by national and local actors. Their similarities arise because they follow their own (private) interests, provide financial resources, are attached to the international scale and hold similar types of activities (e.g., conferences, workshops, etc.). Development agencies, such as USAID, AUSAID and the German GIZ (*Deutsche Gesellschaft fuer internationale Zusammenarbeit*), finance several large government projects and smaller NGO efforts, and they use their financial contribution for policy advocacy and discourse setting. The same is true for business-related international donors, like the Bill & Melinda Gates Foundation and the Ford Foundation. Multilateral organizations, such as branches of the UN offices, implement projects within the frame of international environmental narratives, cooperating with both state and non-state actors. Although divided by their (non)-governmental characteristics, international government agencies and INGOs tend to work together or at least coordinate their work in each country, and thereby form official and unofficial networks of their own.

An interesting case of international actors is the German political foundations, which do not perceive themselves as donor organizations, but as partners for political dialogue. Nevertheless, they bring money into the receiving country but use them for their own political agenda, influencing discourses and policy as an outside actor by cooperating with institutionalized Vietnamese partner organizations. The following section scrutinizes the relations between VNGOs and INGOs and shows how there is, indeed, a certain level of influence that extends from the international to the national realm, and how Vietnamese NGOs are strategic about their connections through their own agenda within the state's governance.

VNGOs and INGOs in Relation

NGOs position themselves and are perceived on different scales. They navigate the translation between the scales using their personal and institutionalized networks that include government institutions, donor agencies and cooperation partners. Analysing the actors and their relation to each other helps in understanding normative ideas, political agendas and their respective position in the system of power. The interactions between actors and across scales create networks that navigate discursive systems (Zink 2013).

A point of interest is the complex connection between VNGOs and INGOs regarding their roles and functions. First, there is no clear divide between the two, apart from their legal registration requirements. Between the two ends of the definitions are Vietnamese staff, the management and directors, and the differing levels of autonomy between the head and country offices. This spectrum has widened as the international organizations recognize the growing capacity of VNGOs and their Vietnamese staff (Hannah 2007). Still, there are several roles and characteristics which focus on power distribution between organizations that do not originate in Vietnam.

In a 2002 report published by the WWF Indochina Programme based in Hanoi (WWF 2002, p. 19), the authors made several recommendations for VNGOs and INGOs to promote sustainable development. According to the report, international organizations are recommended to implement international regimes, support VNGOs in capacity building, act as coordinators of environmental activities between governmental, non-governmental and intergovernmental actors, and support the development of NGOs with participation from VNGOs. Instead of implementing their own agendas, INGOs are perceived as translators of global concerns to the local actors, both within and outside the government. They function more as intermediaries and mediators, rather than as independent actors.

Most INGOs in Vietnam see their responsibility as helping to implement international agreements and global goals, such as the Sustainable Development Goals (SDGs), the Paris Climate Accords and other agreements, and their power as gatekeepers and agenda setters are not to be neglected. INGOs through their capacity building programmes and requirements for funding had an influence on VNGOs (Hannah 2007, pp. 17–18). They also helped to establish local institutionalized organizations with sociocultural and language expertise to implement projects and bring funds into the country. In my interviews with Vietnamese NGO staff, most respondents pointed out that the administrative and finance department was designed according to international donors' regulations. The definition by INGOs matters in outlining the structure and networks of CS and VNGOs.

While INGOs act as an information and service provider from the international to the national scale, VNGOs do the same from the local to the national level. Pham et al. (2010) describe INGOs as network builders for VNGOs, providers of technical expertise and financial resources, and builders in project implementation. As builders, INGOs plan, monitor and evaluate projects. VNGOs have similar functions too. They provide access to communities and non-formalized initiatives. Furthermore, national actors provide expertise in sociocultural knowledge, navigate and arrange contacts with subnational authorities in Vietnam (WWF 2002, p. 184). While INGOs are framed as providing expertise on the international level, VNGOs are experts for the Vietnamese context. Both act as intermediaries and translators between scales and for government institutions. The same dichotomy happens within INGOs too. While international staff are in a leading position providing expertise, Vietnamese personnel are tasked with translating projects for government agencies and communities. They are actively navigating scales, and thereby exercise a framing power hardly seen within NGOs.

Along with the financial budget that INGOs and donors channel into Vietnam come not only accounting and reporting requirements, but also the pressure to succeed. Mosse (2005) explains that development will only be as successful as NGOs' and other actors' need to legitimize their work and their spending. Li (2007) shows in her case study from Indonesia that NGOs and other actors have the will to improve livelihoods, but they do not necessarily do so. The will to improve vanishes behind what Easterly (2014) calls "tyranny of experts" because NGOs tend to use the same environmental narratives—a standard practice of planning, monitoring and evaluation. For VNGOs, this pressure to succeed means that they must prove their successes, deny their shortcomings and failures, and contribute to building narratives to secure funding and cooperation. They also must be moral and conform to values that are embedded in the development projects.

When translating and mediating environmental knowledge and narratives, VNGOs are in a powerful position compared to other CS and non-institutionalized actors, which are mainly active on the local scale. Donors and policymaking bodies perceive VNGOs as representing the ordinary citizens and marginalized groups, as gatekeepers for resources and as entitles for reforms. However, due to their connection to government authorities, VNGOs choose to promote certain aspects of environmental governance (Yasuda 2015). The WWF (2002) report sees NGOs as norm brokers in environmental governance. NGOs can develop better relations with the public by providing services and navigating spaces of participation with the state. They can also navigate spaces with the environment to define actions that are (or not) needed to achieve sustainable development.

Princen and Finger (1994, p. 226) see NGOs' role in environmental action as "increasingly prominent forces in framing environmental issues. They help establish a common language, sometimes, common world views". The WWF (2002) report points out that trainings and workshops should be held to create a common language and understanding of what sustainable development means from an international perspective. NGOs are actors that infuse narratives into the global social governance as shown in Chapter 5.

The NGO scene has changed rapidly in the Vietnamese context over the last decades. The amount of international funding has been continuously reduced by governmental and non-governmental actors in the Global North. As Vietnam has lowered its poverty levels and seeks to become a "developed country" by 2045, Vietnamese organizations are no longer eligible for certain funds dedicated to the least developed or developing countries. The interviewees underlined that the capacity and knowledge of Vietnamese actors have also grown massively, so the international organizations provide funds to local organizations rather than implement work themselves. Some international NGOs have decreased the number of foreign staff in their offices, and, while still uncommon among international organizations, Oxfam, CARE and ActionAid Vietnam are led by Vietnamese country directors.

The localization of international NGO work went hand in hand with an expansion of Vietnamese laws. In recent years, it has become more difficult for international organizations to get permits for activities. This also applies to local organizations with foreign funding. Decree 80/2020/NĐ-CP, which extended the notice needed to hold events, make it harder to amend previously approved projects.

Another change witnessed by the Vietnamese interviewees is in "donor darling topics". For example, in the 2000s it was easy to secure funding from non-Vietnamese donors and NGOs for climate change adaptation projects. But this trend has shifted to projects on plastic and other waste issues; consequently, funds are now allocated to reflect this change. The interviewees also expressed their concerns over the influence that big organizations and donors have on the local organizations. Funding and guidelines are not always adjusted to the local context. The influence of foreign politics on Vietnam could be seen during the Trump administration. The funding cuts under then US President Trump concerning climate change affected donor priorities. Although the international presence is decreasing, the influence from international actors on the agenda in Vietnam remains high.

INGOs' increased capacity and greater autonomy were perceived positively. Within INGOs, funding can be freely allocated to projects within the country office. There are, however, guidelines from head offices that all INGOs must comply. Large, established INGOs, such as Oxfam, WWF and CARE, are funded partly by partnership offices in the Global North and partly by special project funds. For example, in wildlife conservation, the specific budget allocation means receiving extra project funds for a certain species (e.g., tigers, bears), even if these species are not part of Vietnam's conservation plan. Apart from the special funds, international NGO staff said that the autonomy they received from the head offices gave them the decision-making power to adjust international strategies to the national context.

In the organizations where international staff fill the management levels, the non-Vietnamese team members value working with staff who have been "internationalized", this includes those who have previously studied or worked abroad. Similarly, Vietnamese staff value working with foreign staff who have been "contextualized locally". The latter is not much of a priority as only a small minority of foreign staff choose to learn Vietnamese or seek knowledge of the local sociopolitical or cultural contexts. Local colleagues usually fill the knowledge gaps, or they help when a foreign staff member is unable to handle a situation in a locally appropriate manner. When foreign staff spoke Vietnamese, their language skills helped them to communicate effectively and to build trust, making the relationship informal, yet professional.

Besides agenda-setting within NGOs, the Vietnamese government also has an influence on NGOs. The government states the areas for NGOs to work on, and this influences the work of VNGOs as their priorities often shape the funding of foreign development aid agencies. Although NGOs welcome foreign experts to share their skills and knowledge, they prefer, however, that foreign experts spend time in Vietnam and be willing to adjust concepts to the local context. These conceptual frameworks are used to negotiate tensions between environment and economy, evolving around a new type of consumerism that seeks to be sustainable.

Some interviewees see an issue with Western people living in Vietnam for a short period because these Westerners usually have a superficial perception of the environmental problems in Vietnam. A good example is the waste problem. All the Westerners I have interviewed for this research mentioned waste management as an environmental problem in Vietnam. Although Vietnamese interviewees also mentioned waste management as an issue, they see the problem as more of a sociopolitical matter rather than a cultural one. They attributed the problem to waste exported by European countries to Vietnam, a lack of waste management system and a rapid development towards consumerism. The foreign interviewees, however, deemed the problem as a lack of public awareness and ignorance. This shows that foreign experts localized problems to the extent that they are disconnected from global issues. Hence, this underlines a problem with adapting to Vietnamese society and understanding how politics in the country works.

The high turnover of foreign staff is a problem, with many of them staying in Vietnam for only three years. Hence, their knowledge of the Vietnamese sociocultural context and working style is very limited. Some international team members undergo a few days of training before they arrive in Vietnam or within the first few weeks in the country. Most cultural learning takes places informally through work and in their personal life. The lifestyles of foreign experts are perceived by the locals as privileged, with high salaries and the prerogative to live above local laws. "Here, foreigners can drive over a red light, and no one cares", said one informant. This shows the perception of hierarchy in power relations.

Foreign NGO staff remain deeply entrenched in colonial and racist perceptions. A white European interviewee shared his view on the role of nature in Vietnam: "It is hard to experience nature—while in Germany, there are many culture landscapes, you don't have too many of them here. There is a missing tradition of nature protection areas and a functional relationship to nature and environment exists. A lot is eaten, a lot is hunted down" (Interview by the author). This statement illustrates othering and stereotypes from a white perspective. First, there is discord between experiencing nature, which was earlier defined as non-human, and exemplifying nature in a German context with "*Kulturlandschaften*", which is how humans interact with the environment. Moreover, the interviewee blamed the functional relationship to nature in Vietnam as the reason why it is difficult to experience nature in a German sense. But *Kulturlandschaften* refers to an environment where nature produces food for humans. So why is it hard to experience nature when everything is hunted down and eaten in Vietnam, but easy to do so when the same is happening in Germany? Lastly, his criticism of a lack of nature protection areas not only overlooks the number of national parks that exist in Vietnam, but also does not reflect on its colonial history. Similar opinions were expressed by non-Vietnamese NGO staff who have been working in Vietnam for a short time; those who have lived in Vietnam for a longer period and who have established personal connections (such as having a Vietnamese spouse) seemed to be more reflective on this matter. This shows how much othering, stereotypes and racism are prevalent in the international NGO working environment.

In contrast, a few Vietnamese interviewees openly criticized the "foreign" attitude towards nature and the environment: "Foreigners see money when looking at trees, we see nature." This statement refers to conflicting interests in some areas of Vietnam, where foreign-financed development projects conflict with local community projects, and the ecological modernization programme is openly criticized. Still, the interviewees suggested that there is no need to abolish the Sustainable Development narrative. Instead, concerned actors must jointly define its meaning to include wars, spirits, ecological homes and beliefs. An interviewee added that the Vietnamese mainstream society should not define civilized vs backward per se and instead negotiate for a more inclusive understanding of the environment. This implies that nature relationships are not just management of nature; they are much more multidimensional.

In my interviews, stereotypes are evident not only from the international to the national level, but also from the national to the community contexts. The interviewees described rural areas and its inhabitants as more natural than their urban counterparts. They also had certain notions regarding people living in rural and urban landscapes, such as describing rural people as hardworking and city dwellers as creative. Similarly, Laos was seen as more natural, where the people were living much closer to nature. These perceptions were derived from colonial narratives in "Indochina". Weller (2006, p. 5) observes the shift in meaning of urbanized people living close to nature, seeing the development of infrastructure as "another victory for humanity". But at the same time, the people have greater environmental awareness that is distant to their livelihoods, yet function as an aspiration for them. In the process of redefining the urban human-nature relations, maintaining the rural surroundings becomes both a goal to strive for and an obstacle to modernization (Schwenkel 2020, p. 10).

Rural Vietnam (*quê*) is depicted as a home, a place of belonging, and a location that needs to be developed. It reconciles the images of well-being with backwardness and hardship at the same time. In her book on care work in Vietnam, Nguyen (2014) described *quê* as a place of the past and not the future. Agriculture prevails in the countryside, and the people are not necessarily integrated into the market economy. This means reforming behaviours and mindset of the rural population and shifting the norms and values of the state's governance away from its political centre. Framing existing behaviour patterns and traditions as backward and uncivilized helps to aspire developments in the countryside. At the end of this chapter, I will narrate an account of a community that is part of the *nông thôn mới* policy programme to build a modernized countryside. The case study depicts why rural citizens are grateful for the state's efforts, yet remain critical of them.

While international actors are funding local NGOs instead of being present in the country themselves, funding regulations are perceived to hinder Vietnamese NGOs from implementing projects in a suitable way for the local conditions. Projects that are seen as successful by local NGOs have to be cancelled or cut short due to difficulty and lack of capacity in meeting accounting requirements, especially for smaller NGOs. Hence, the field of NGOs in Vietnam is often dominated by a few big players. The knowledge and the resources needed to handle donor requirements keep smaller players that are not acquainted with international working environments out of the game, thereby restricting NGO work.

On the positive side, INGOs and donor organizations help Vietnamese NGOs extend their network. This fulfils Vietnam's vision of becoming part of the global scale that extend beyond national interests—a vision shared by both governmental and non-governmental actors. Vietnamese NGOs perceive networks as necessary for cooperation with international actors, both within ASEAN and beyond, because environmental problems cannot be solved within one country, and networks are needed to share experiences and research on transnational issues, such as climate change, water flows and wildlife trade.

Furthermore, INGOs and international frameworks and agreements provide the space for VNGOs to work within the sensitive political context. As outsiders to Vietnamese society and are not accustomed to unspoken rules, INGOs can easily feign ignorance. Until 2020, INGOs were allowed to hold meetings without permission. However, this is no longer possible after a change in the law regulating Oversea Development Assistance (Decree 56/2020/ND-CP). It remains to be seen how strictly this regulation will be implemented. One INGO reported that they received a warning after they held a meeting without permission at their office. The role of INGOs is under scrutiny, and it is unclear how far the Vietnamese government will hold them accountable without interfering in their work.

Generally, all NGOs follow a multi-level, multi-stakeholder framework. This means that they work with communities and the local government, researchers and the international network, and universities and ethnic minorities on the same issue. They hope to induce change by simultaneously working with different people on different scales. By working on the local, national and international levels, they connect the scales to each other and emphasize the mutual flows between them. However, the authoritarian political system restricts the opportunities from international donors, which play a crucial role in making space for VNGOs. The second vignette demonstrates the negotiations between actors and scales and the importance of following the processes.

Case Study 2: Promoting Lifestyle Changes in Southern Central Vietnam

Most ceremonies at universities in Vietnam consist of female student dancers in *áo dài*, a student boy band and official university representatives giving speeches emphasizing on partnership, solidarity and hierarchies. The one I was attending was no exception—it was an opening ceremony for an educational environmental week that was organized by NGO 2 and a university and funded by a German public organization.

The vice rector of the university gave the opening speech. He welcomed everyone and introduced the history and mission of the school. He underlined the importance of Sustainable Development to achieve an environment that is "*xanh hơn, sạch hơn, đẹp hơn*". Next, the director of the funding organization introduced his organization and outlined the concept of social-ecological transformation (SET). This term is translated into both Sustainable Development and "*chuyển dịch sinh thái xã hội*" at different points of his speech. It is evident that the international cooperation partner tried to push for a different framing and a systemic critique of ecological crises. The fact that the term "social-ecological transformation" is interchangeably used with Sustainable Development is evidence of the NGO navigating between the aspirations of the financing organization and the political and cultural contexts that it is operating in.

The educational week is organized by the NGO during the summer of each year. The NGO works with different universities across the country on different topics to promote good environmental behaviour and a "green lifestyle" in university students. The theme for the week was eco-footprint of tourism. As with the other educational projects, a core group of students were trained in environmental topics, making them leaders to drive positive changes in their communities.

NGO 2 is an organization established in the late 2000s. The organization has six full-time and six part-time staff. It gets most of its funding from multilateral and governmental development aid organizations, and it focuses mainly on environmental education for students. Its projects include conducting trainings and workshops in universities, researching and raising awareness of environmental behaviour, and publishing educational materials on food safety, plastic waste and sustainable consumption. This is in line with the organization's vision to have different societal groups actively engaging in protecting the environment and managing natural resources for a sustainable life. NGO 2 has a strong emphasis on environmental stewardship and sustainability, and this is reflected not only in their vision statement, but also in their publications and programmes. Sustainable Development is connected to an individual's environmental behaviour through the concepts of footprint and green consumerism.

The ceremony kicked off a day-long programme for all interested students. A fair was held to sell eco-friendly products, and at the same time posters and videos were used to educate students on wasteful consumption. The students were introduced to homegrown organic materials and their traditional uses, for example, banana leaves for food packaging and bamboo for making furniture.

One of the sellers for organic cosmetic products was a professor of engineering at the university. She used her passion in environmental behaviour to produce and sell organic cosmetics as a side hustle to supplement her low salary as a university staff. Thereby, she has become what Nguyen (2018) has coined "moral citizen" under the Vietnamese government's "socialization" process. The state renders socialism and capitalism compatible by having values and norms which are centred on citizenship. The government aims to develop the market economy through self-optimization and self-entrepreneurship while building a prosperous, socialist nation.

Another part of the fair displayed visual effects of global warming, deforestation and marine habitat destruction. A documentary on a campaign called "Clean Up World" was shown. This campaign has been implemented in many countries globally. Participants in the campaign regularly headed to beaches for cleanup. Hence, the responsibility for the common good has been decentralized to society actors, without the state having to intervene in production and consumption patterns. This means that state actors do not have to challenge the system of non-sustainable economic growth that legitimizes the CPV government's rule.

Another activity focused on reducing and recycling plastic waste. The audience was led from the lecture hall to another university building, where a mosaic mural made from plastic bottles was uncovered. The mural depicted a whale whose tail was a rainbow. This wall was chosen because it was a place frequented regularly by most students, therefore serving as a daily reminder to "reduce, reuse and recycle". An NGO representative added that this mural reminded the students that waste was not merely garbage but also a resource, in this case, a resource for creating art.

Other activities focused on team building and raising environmental awareness. When I spoke to the students watching an electric motorboat racing, no one seemed to link the activity to environmental concerns but rather enjoyed it for entertainment only.

Inside the lecture hall, groups of students gathered to design t-shirts surrounding the message of "reduce, reuse and recycle". The winner with the best design would win a prize. The activities were organized by the student volunteers with the support of their lecturers from the university. The students were given as much autonomy as possible in organizing the opening day, so that they could learn about environmental topics. This approach also helped to break down hierarchies and create a sense of ownership among the students.

Another important feature of the opening day was the selfie spots. Outside of the lecture hall, the students gathered in front of a large backdrop to take pictures. Next to the backdrop was a large sheet and containers of paint. The students were invited to leave their handprints on the sheet as a sign of commitment and participation in the project. Many students took selfies and posted online. Hence, the project was taking place not only on campus, but also in the digital space.

Both online and offline, the students were more interested in making friends than in learning about environmental topics. This became obvious when I asked the students about their motivation for participating in the programme. All of them replied that they participated because they wanted to spend time with their friends and establish new friendships. Although only a few students mentioned that they wanted to learn about environmental topics, nevertheless all the students participated actively in the activities. They are what NGOs call an ideal group for raising awareness around environmental problems.

When I talked to the students about their perception of the environment, it became apparent that they were aware about changes in the environment. The students noted that summers had turned unusually hot and dry, and they linked this to climate change. They also mentioned that the huge influx of tourists had negative consequences on the environment. All of them said that they had taken part in environment-related activities, such as the clean-up campaigns. When they were asked what Sustainable Development means to them, no one felt confident to reply. One group of students felt that Sustainable Development should be taught at the Faculty for Environment and that only the students there were familiar with the concept. They understood Sustainable Development to be both academic and technical.

The students' understanding of the environment and the waste problem were obvious from the results of the t-shirt painting competition. The recurring themes drawn on the t-shirts were human behaviour causing devastating environmental consequences and their possible solutions. One t-shirt, for example, depicted the recycling symbol, a bug, a wind turbine, an electric car, a plug and crossed-out plastic bags as a loose collage of visuals. Another showed a human eating a fish, which then eats a plastic bag. Yet, another t-shirt showcased a tree framed by a bicycle, a car, wind turbines, solar power and electricity poles. The last t-shirt showed a divided globe, with one half completely destroyed and the other half green and thriving. The student who drew this t-shirt explained that the second half was occupied by "people with awareness" ("*người có ý thức*"), which he symbolized using images of technology, giving the general notion that "humans" are to blame for environmental crises.

A group of female students designed a unique t-shirt. The t-shirt depicted a woman whose hair was made of plants and her clothes made of feathers, and she was surrounded by fire and an elephant. This t-shirt was different from the others because it did not show environmental devastation and technological solutions. Instead, it made a human-nature connection and that humans are part of the environment. The feminine figure refers to the femininity of nature and eco-feminism. Here, gender is linked to the environment.

The t-shirt designs showed that the government had been successful in instilling a sense of environmental responsibility in the students. The students were aware that their behaviour played a crucial role in overcoming environmental challenges.

The first day concluded with three lectures given by external guests. One lecture touched on sustainable energy models for the communities; another lecture talked about preserving the culture of ethnic minorities and the influence of tourism; the last lecture was about responsible consumption.

The lecture on ethnic minority and tourism was conducted by an ethnic minority—a rare representation. Although the NGO strived to integrate ethnic minorities' perspectives into their projects, there has been no representation from this community in their publications. The NGO was aware of issues of representation in the programme. Hence, it invited three external facilitators from North, Central and South Vietnam to coordinate the programme, moderate talks and organize games. This provided the students with a diversity of perspectives. The NGO also strived to maintain a gender balance among the speakers and participants. Thus, it used representation as a tool to overcome unequal power relations.

As the lectures were held concurrently, I chose to attend the lecture on sustainable energy models. The presenter made the lecture interactive by quizzing the students and using a narrative style to illustrate her points. By engaging with the students, she broke the pattern of authority typically present in a university classroom. The presenter's main message was that Vietnam needs to reduce its reliance on fossil energy, especially coal, and switch to renewable energy instead. This can be done by decentralizing the energy system and reducing energy usage. The latter stands in stark contrast to the Vietnamese government's narrative that an increase in energy usage is necessary for the country's development. To reduce energy usage, *"công dân sinh thái"* ("eco-citizens") are needed, which means that individual consumption needs to be reduced to a moderate level. The presenter used a pyramid diagram to summarize her argument. At the basis of the pyramid is ecology, in the middle is society and at the top is economy: "The problem is that if ecology is destroyed, then we also do not have culture or economy." This illustrates the interconnectedness in environmental issues, and the whole society is responsible for solutions. This means the government and the people are important actors for change.

On the second day, only pre-selected students were allowed to join the project, and they were divided into groups according to their participation in the fieldwork programme. Part of the project was a two-day excursion either to an island to work with the local community on island tourism and marine waste, or to a farm that used ecological agriculture. I joined the second group.

On day two, the NGO promoted an ecological lifestyle, which was defined as a "way to live to protect the environment, including the society and economy", making environment the space of interaction for several actors. According to the NGO, an ecological lifestyle is the same as a sustainable lifestyle. This lifestyle was displayed through different activities at the week-long programme.

The first activity was the footprint exercise, which calculates an individual's ecological footprint based on the demand an individual made on natural resources. This exercise depicts an individual's role in defining problems and finding solutions within the current system (McManus and Haughton 2006).

Using a smartphone app, we calculated our footprints and shared the results. While socioeconomic factors had an impact on an individual's footprint, the students were not guided on how their lifestyles impacted their amount of footprint. For example, the students reported different amounts of footprint due to their different lifestyles: some lived with their parents in the rural areas and commuted to school by motorbike every day; others went on frequent overseas trips. Nevertheless, the students were advised to reduce their footprints. This illustrates how individuals can be transformed into moral citizens for environmental sustainability.

To identify pro-environmental behaviour, the students were divided into groups to brainstorm ideas to save water and reduce waste. They came up with ideas such as "green coffee" using local products and without producing any waste, but they were harshly criticized on the feasibility and originality of these ideas. The discussion instead turned to praising Germany's adoption of e-mobility. Here the "green" Global North's eco-imperial narrative laid out perceptions and power structures on the international scale. Germany was repeatedly framed as a role model for an ecological lifestyle and technical solutions for environmental problems, although it produces large footprint and lacks in e-mobility technology.

Interactive activities, including a discussion on the pros and cons of using plastic bags, took place along other sessions, which helped to establish cross-actor connections. In the first lecture, a professor from the university explained the concept of ecological agriculture (*nông nghiệp sinh thái*). He showed a movie on sustainable land use in Brazil, France and Ukraine, and used concepts of nature to limit human involvement in the agricultural field. Technology was shown as devastating to the environment which contradicted with the earlier programmes where technology was seen to offer solutions for environmental crises. Ecological agriculture was characterized as "safe, sustainable and healthy". For this concept, the NGO adopted alternative narratives, but the trainers and students saw no contradictions with other concepts. To them, there are various practices that can be implemented alongside each other.

In the concluding lecture of day two, the facilitators jointly explained the project planning process commonly used in the INGO field: determining the project output and outcome, discussing the theory of change, determining the target audience, designing the project, setting the timeline and planning the budget. This introduction served to prepare the students in planning projects themselves—a desired outcome for the next two days. The students were confused over the technicalities of *mục tiêu* and *mục đích*, what I would translate as output and outcome in this context, because the facilitators failed to adjust the logical framework contexts familiar to them. With this comes an understanding of a hierarchy in project planning: who determines the concept and for whom?

The next day we gathered early on campus to board the buses for the field trip. The journey to the farm, a place for ecological agriculture, took about two hours from the city. The atmosphere on the bus was lively, and the students took turns to sing on the microphone. After some detours, we finally arrived at the farm. The sun was already burning hot, so we took refuge at a shaded place and gathered around some tables. The founder of the farm welcomed and served us some drinks made from herbs. As we sipped on our drinks, he explained the story behind the farm. The founder revealed that he had previously worked at an overseas permaculture farm, where he was inspired to introduce the farming techniques that he learnt to Vietnam. His idea was to set up a farm stay to generate income from ecotourism and to grow native plant species using "*mô hình tự nhiên*", a natural model. His philosophy was to link his farm with nature.

His plan was initially met with obstacles, some of which still exist today. Originally, the farm was to be established in the Central Highlands. However, the local authorities would not grant permission for the project. As a result, the farm moved to its current location, but problems with the local authorities persisted with the Ministry for Tourism, Biotechnology and Environment maintaining much authority.

We were brought on a short tour around the farm to see vegetables, herbs, flowers, chickens and sheep. Next, we were divided into groups to compete in two games. In the first game, we competed in how fast we could plant herb seedlings. In the second game, we were brought to the chicken farm where we competed in how fast we could collect the eggs. I am not sure how the chickens perceive a bunch of nervous strangers running around their home. I am also doubtful of the effectiveness of this approach in connecting nature with competitive games.

The same thoughts remained when two more games were continued after lunch. As it was far to walk to the sunflower fields, we were driven over by trucks. At the sunflower fields, we competed in how fast we could plant sunflower seeds. Next, we were brought to the sheep barn and our goal was to take the best group photo with the lambs. However, the lambs seemed neither amused nor willing to pose for pictures; instead, they seemed rather terrified by attempts to catch them and took photos in cute poses. The objective of the sheep photo competition was to connect the students' real experiences with nature. But none of the students gained any knowledge about sheep from this game. Instead, they perceived the animal as a prop for their social media presence. The organizers and the farm owner did not provide any narrative and context; therefore, nature became an object rather than a subject of interest.

Day two of the field trip was dedicated to developing project ideas. The students gathered around the farm to brainstorm ideas on reducing ecological footprint for their upcoming projects. They had to give a presentation on the final day of the programme. The winners could use the prize money to implement their projects.

The last day of the programme kicked off with two concurrent talks by invited guests. The guests discussed the consequences of economic development for the environment and deemed state policies as well as individual actions as keys to achieving changes. In front of their peers and the jury, the students pitched their project proposals. Their ideas varied widely. I mentioned three projects in detail below to underline how environment is perceived socially and is connected to power relations.

One group, which visited the island for the field trip, suggested forming a volunteer group that would travel regularly to the island to teach English to the people living there. The students argued that a knowledge of the English language would have an "influence on the eco-system by the people, change the understanding of the people to develop economy, society and environment". They hoped that with improved English language skills, the people living on the island would be able to take ownership and be empowered to influence the tourism businesses there. They would be less reliant on exterior agents and tourist companies. The students also saw it necessary to change the people's mindset on economic development. In this pitch, environmental concerns were clearly secondary, instead the focus was on developing the economy.

The second group proposed a project to teach children living at the local SOS village on saving water and electricity. The students argued that "they have nothing, they do not have a chance and should also learn what everyone knows about saving water and electricity". Here, the students assumed that the children lacked an awareness and knowledge of the environment due to their poor family background and low socioeconomic status. In fact, the children living at the SOS village attend regular schools in Vietnam, and they are part of the national education system. The students also did not reflect on their own water and electricity consumption; they assumed this to be the same as those children at the SOS village, although it is questionable to compare the ecological footprints produced by different living situations. Thus, environment helps to enlighten the socially "other" on their top-down perspectives.

The third group suggested allocating a "green apartment" on the university campus. This apartment serves as an ecological space, which includes a student dormitory so that the students do not have to commute to school daily. This helps to reduce their carbon footprint and save their time and improve their health. In addition, there is a book corner in the apartment for students to exchange books, and a café which does not serve plastic cups. This proposal makes a win-win connection between social actors and the environment. It links different concerns together and makes the project less hierarchical than the other two by seeing the institutional actors as responsible for change.

Through this third pitch, it is evident that the students and NGO staff internalized ecological modernization through self-optimization and self-entrepreneurship. Their actions for environmental sustainability are possible within an authoritarian framework that combines a socialist system with a market-based economy. Although technical solutions prevail, the socialist system is crucial in maintaining a green and prosperous nation to which all citizens contribute through their individual actions. Consumption, production and environmental actions contribute not only to an individual's accumulation of wealth but also to a common environmental goal. Thus, sustainable development takes a "moral turn", even in processes across scales (Derks and Nguyen 2020).

Conclusion: The Need to Follow Processes across Scales

This chapter has widened our perspective beyond the Vietnamese context to the environmental narratives and actors in it. It points out that we must not only look at the global and the national levels, but also draw conclusions from both levels to see the influences that the actors have on each other. Power relations is not equalized with scales: the global scale is not necessarily more powerful than the national scale, and INGOs do not necessarily make decisions in the NGO scene in Vietnam. It is apparent in the case study that ecological modernization has been successful, but nevertheless there are contradictions in the NGO projects. These contradictions might not be addressed and identified, but the donor's narrative exemplifies a need to identify problems and find solutions. Disregarding the contradictions would mean ignoring history, context and the agendas that have been central to forming socioecological realities across the world. Now that we have a better insight into cross-scale processes and their interconnectedness, we can analyse the universal narrative of Sustainable Development within the Vietnamese context. This leads us to the final argument why Sustainable Development is so central and powerful for environmental rule, even amid challenges.

5

The Narrative of Sustainable Development in Vietnam: Why It Is So Powerful

As shown in the previous chapter, the Sustainable Development narrative has been on a global conceptual landslide over the last three decades. It was derived for the global scale but has been adapted to multiple contexts within the ecological modernization programme. This chapter examines how the narrative is used by the state and non-governmental organizations (NGOs) for governance in Vietnam, and which other alternatives exist. The Sustainable Development narrative has been so successful because it offers space for adaptation and is not perceived as a threat to power relations. The Vietnamese government uses it to further extend its environmental governance. Furthermore, the Sustainable Development narrative allows for ecological modernization, embraces the capitalist economic model and sustains economic growth. By not questioning the current governance system in Vietnam, this narrative has a competitive advantage compared to other narratives. As a result, NGOs have used the Sustainable Development narrative to maintain their place between the Vietnamese government and international funding organizations. Their theories of change are limited to the conceptual path dependency and their perception of the environment, with the identification of environmental crises as part of it. They use Sustainable Development as a narrative to serve power relations but leaves space for manoeuvre when it comes to practical implementation.

The Vietnamese Policy Adaptation of Sustainable Development

The 1980s were a time of transformation in Vietnam. Often attributed to the 6th National Congress of the Communist Party of Vietnam (CPV) in 1986, the *Đổi Mới* renovation politics had actually begun much earlier. In that year, famine and poverty forced people to transform policies (Kerkvliet 2005). For example, family units adopted collective farming, following earlier village traditions, instead of the cooperative structure. Another reason for the major change in politics was the global threat to socialism. As the Soviet Union and East Germany began to crumble and eventually fell apart, the CPV recognized that if they wanted to hold on to power, they might need to change their way of governance. Among the officially proclaimed changes was the introduction of a "market economy in socialist direction" (*kinh tế thị trường định hướng xã hội chủ nghĩa*). This meant the return of private property and the welcoming of foreign organizations into Vietnam. It led to a new era in Vietnam's environmental policy under the Sustainable Development narrative.

A few publications on environmental policies in Vietnam have identified the political path and rhetoric of market orientation. In his paper on the Vietnamese climate change strategy, Fortier (2010) notes that the strategy is characterized by the "ecological modernization" discourse. He argues further that growth is the top priority of the climate change strategy, which is itself firmly rooted in the Sustainable Development narrative. To Fortier, environmental governance enables economic growth with the state and its organs having a central role in the management of climate crises, but at the same time decentralizing policy implementation through a variety of actors. It maintains the state's power despite the reallocation of responsibilities to non-state actors, and it reinforces hegemony. Cole and Ingalls (2020) describe how the Green Growth narrative has tried to combine economic development with environmental concerns and is supported by donor organizations, but the policies under this framework have not reduced environmental harm. Despite the crucial role of donor organizations and international organizations in implementing the Sustainable Development narrative, local elites also play a role in reproducing discourses as shown in the following analysis of policy papers. The West and the Rest argument that creates a dichotomy between hegemonic countries in the Western hemisphere and formerly colonized countries does not hold for the Vietnamese case. McElwee (2012) emphasizes that the market-socialist state brings together a strong government that promotes market-based solutions for environmental problems, similar to payments for environmental services. Like in socialist China, privatization in Vietnam has created a "socialism from afar" (Ong and Li 2008). Through capitalist conducts like self-entrepreneurship and self-optimization, citizens and small- to medium-scale businesses act independently but in the state's interest. At the same time, the state promotes socialist goals and a sense of unity alongside market values and policies of commodification and privatization. How the Vietnamese state defines and uses the Sustainable Development paradigm for forming its governance becomes evident when analysing the discourse used in its policymaking.

The term "Sustainable Development" and the rise of the discourse coincided with the beginning of Vietnam's renovation politics in 1986. The national policy at that time was summarized as *phát triển kinh tế-xã hội* (literally, economic-social development, but translated by the Vietnamese government to socioeconomic development) and remains so today. Sustainable Development (in Vietnamese, *phát triển bền vững*) complemented socioeconomic development from the 1990s onwards. The relation between the two paradigms is still not clear today, but a statement from the commission tasked with reforming the environmental protection law in 2020 suggested integrating both paradigms into "economic-social development in a sustainable way"[1] (Working Group for the Reform of the Law for Environmental Protection 2020, p. 7). Translated literally, *phát triển bền vững* has a slightly different connotation from its English term. *Bền* means durable and long-lasting and *Vững* can be translated as strong and unshaken, thus characterizing the kind of development as a positive, stable and active outlook into the future. This is the first evidence of how the Sustainable Development discourse has been integrated and adapted into the Vietnamese context.

---

1. *"phát triển kinh tế - xã hội cách bền vững"*.

Several government publications explicitly define Sustainable Development by using definitions and concepts from UN processes and frameworks. For example, Vietnam Agenda 21, a UN initiated document in 2004, quoted the Brundtland definition (translated into Vietnamese) and elaborated the term Sustainable Development from the 1992 summit. This is repeated in various versions of the Vietnam Law for Environmental Protection from 2005 and 2014.

In 2012, the Vietnamese government published the report "Implementation of Sustainable Development: Report at the United Nations Conference on Sustainable Development (RIO+20)", which summarizes the Vietnamese path to Sustainable Development over twenty years of implementation and strategic orientation. The report states that there have been major gains across all three pillars (economic, social and environmental), including in economic growth and improvement in the "population's material and spiritual living standards" (Socialist Republic of Vietnam 2012, Foreword). It describes Sustainable Development as "a common trend along which the entire humankind is endeavouring" (Socialist Republic of Vietnam 2012, Foreword) and "an inevitable path and will be vividly and effectively realized in Viet Nam's process of development and integration. It is also an important strategic goal that the Communist Party, Government and people of Viet Nam are determined to attain" (Socialist Republic of Vietnam 2012, Foreword). The report depicts Green Economy as a way to achieve Sustainable Development. Other key terms in the report are modernization (*hiện đại hóa*), industrialization (*công nghiệp hóa*) and natural resources (*tài nguyên thiên nhiên*). Forests, water and land are described as natural resources that the people "manage, exploit, use" ("*được quản lý, khai thác,*[2] *sử dụng*") (Central Committee of the Communist Party of Vietnam 2013), and their worth is measured in relation to the Gross Domestic Product (GDP). Generally, natural science studies of the environment's degradation and its economic potential are keys to identifying problems and finding solutions. Almost every paper suggests investing in science (*khoa học*), technology (*công nghệ*) and research. Technological fixes are another characteristic of the Green Economy (Fatheuer, Fuhr, and Unmüssig 2015), and they are used as a scientific approach to achieve ecological modernization (Mol 2003).

---

2. The translation of *khai thác* is more complex than the English term "to exploit". In Vietnamese, it has a positive connotation and includes making investment to develop something.

Technological fixes have been widely applied by the Vietnamese state. For example, the development of the Mekong Delta was based on scientific research and technology, sidelining sociocultural livelihood concerns and alternative forms of information (Benedikter 2014, pp. 270–86). Technology is used on large-scale infrastructure projects to increase the effectiveness and efficiency of trade and value chain operations. In climate change, the government acknowledged that scientific knowledge needs to be accompanied by "traditional experience and traditional knowledge", but it does not provide any detail or publishes any paper on narratives and concepts that are beyond the modern scientific understanding (Lindegaard 2020, p. 166). Different forms of knowledge are also used in the state-building process to include ethnic minorities.

In the 2000s and 2010s, in line with the international processes, key terms such as Green Economy and Green Growth entered the Vietnamese government's official vocabulary. Green Growth is linked to Sustainable Development: it frames economic gains and environmental protection as mutually inclusive since a system change is not envisioned under the Sustainable Development paradigm. It is loosely based on the concepts around Green Economy that was introduced in a report written by three environmental economists in 1989, and that was further developed into a book in 2000 (Pearce, Markandya and Barbier 2000). The report and book provide practical ideas to implement the Sustainable Development concept. They acknowledge the need for environmental action without having to revolutionize the political-economic system, thereby sticking to the capitalist narrative that won the Cold War.

The understanding of what Sustainable Development means in the Vietnamese context is also closely connected to the Green Growth concept. The 2012 Green Growth strategy states:

> Green Growth is an important part of sustainable development, ensuring fast, efficient, sustainable economic development and contributing an important part towards implementing the National Climate Change Strategy.
>
> Green Growth has to come from the people and be for the people, contribute to the creation of jobs, eradicate hunger and reduce poverty, improve the living standard and livelihood of citizens.
>
> Green Growth is based on investment to preserve, develop and use efficiently natural capital, reduce Greenhouse Gas Emissions as well as improve the quality of environment by stimulating economic growth. Green Growth has to be based on research and modern technology, appropriate

for the Vietnamese context[3] (Office of the Prime Minister of the Socialist Republic of Vietnam 2012c, p. 1).

As we can see, Green Growth is adapted as a key strategy combined with the development paradigm and other economic terms like investment and natural capital (*đầu tư, nguồn vốn tự nhiên*), and it uses "greening" (*xanh hóa*) production and consumerism to alleviate poverty (*xóa đói giảm nghèo*).

The term is later picked up and reproduced in other papers. For example,

> Push for a change in the growth model, close to the restructuring of the economy according to green growth and sustainable development. Issue a set of indicators to evaluate the outcomes sustainable development, green growth introduced into the national policy criteria; a pilot for the development of a green economy, green industry, green urban areas, green rural areas[4] (Central Committee of the Communist Party of Vietnam 2013).

The "green" concept overlaps with national development, such as in housing policy and urban planning. For example, Harms (2012) describes the recent phenomena of the "*xanh, sạch, đẹp*" policy in urban development in which the citizens in Ho Chi Minh City accepted displacement for the greater good of a "green, clean, beautiful" city. This slogan has become visible all around the country: smaller towns and bigger cities display it along the streets and garbage trucks have it printed on the side. The modern urban space incorporates the "green" branding in a clean, beautiful and urban way. Thus, making the environment the central part of the transformation process.

This incorporation into policy planning in general, and in urban planning specifically, is not new and has been integrated into different systems.

---

3. "*Tăng trưởng xanh là một nội dung quan trọng của phát triển bền vững, đảm bảo phát triển kinh tế nhanh, hiệu quả, bền vững và góp phần quan trọng thực hiện Chiến lược quốc gia về biến đổi khí hậu.*
   Tăng trưởng xanh phải do con người và vì con người, góp phần tạo việc làm, xóa đói giảm nghèo, nâng cao đời sống vật chất và tinh thần của người dân.
   Tăng trưởng xanh dựa trên tăng cường đầu tư vào bảo tồn, phát triển và sử dụng hiệu quả các nguồn vốn tự nhiên, giảm phát thải khí nhà kính, cải thiện nâng cao chất lượng môi trường, qua đó kích thích tăng trưởng kinh tế.
   Tăng trưởng xanh phải dựa trên cơ sở khoa học và công nghệ hiện đại, phù hợp với điều kiện Việt Nam."

4. "*Thúc đẩy chuyển đổi mô hình tăng trưởng gắn với cơ cấu lại nền kinh tế theo hướng tăng trưởng xanh và phát triển bền vững. Ban hành bộ chỉ số đánh giá kết quả phát triển bền vững, tăng trưởng xanh đưa vào bộ tiêu chí quốc gia; thí điểm phát triển mô hình kinh tế xanh, công nghiệp xanh, đô thị xanh, nông thôn xanh.*"

Schwenkel (2017; 2020) argues that the environment and ecology of urban spaces have played a role as "urban eco-socialist ideals" in the state narrative, connecting the idea of progress with the environment. In contemporary Vietnam, the "greening" process is still going strong and is closely connected to the term "eco" (*sinh thái*) (Schwenkel 2017), which spread across cities (*thành phố sinh thái*), buildings (*công trình sinh thái*), tourism (*du lịch sinh thái*) and so on. But different to the eco-socialist idea and the pre-*Đổi Mới* era, the eco and green concepts are adapted to a different context of progress, with prosperity being the main indicator. Achieving prosperity becomes the responsibility of the individual and economic actors rather than the state's duty. Businesses have consequently taken over the narratives to market their products, and so have NGOs in framing individual lifestyle choices. The socialist idea of progress continues to matter today, but with different actors involved, and the state takes a step back through deregulation.

Throughout the policy papers today, we see the idea of "green" evolving alongside a shift in international influence across sectors. For example, the definition of "green jobs" ("*việc làm xanh*") has been adapted directly by the UNEP:

> Green jobs: Is the work in industry, production, research and development, general administration and service, that take part in preserving, restoring environmental quality. In detail, but not exclusively, this can mean work that helps to protect ecosystems and biodiversity, decreases consumption of energy, material, water by strategies with high efficiency, reduce carbon emissions of the economy and reduce or eliminate all forms of waste and pollution (UNEP)[5] (Office of the Prime Minister of the Socialist Republic of Vietnam 2012c).

The green economy nexus also comes in the form of sustainable consumerism or green consumerism. Being "green" has mattered in Vietnam for decades, but its framing and connected concepts and implementation have evolved together with the changing international discourses that Vietnam has adapted for itself.

---

5. *Việc làm xanh: Là những công việc trong nông nghiệp, sản xuất, nghiên cứu và phát triển, hoạt động hành chính, và dịch vụ đóng góp đáng kể để bảo tồn, khôi phục lại chất lượng môi trường. Cụ thể, nhưng không loại trừ, là công việc giúp bảo vệ hệ sinh thái và đa dạng sinh học, giảm tiêu thụ năng lượng, vật liệu, và nước thông qua các chiến lược hiệu quả cao, giảm phát thải các-bon cho nền kinh tế và giảm thiểu hoặc tránh hoàn toàn tất cả các hình thức chất thải và ô nhiễm (UNEP).*

The discourse of Sustainable Development is not without frictions, however. Despite the abovementioned narratives surrounding Sustainable Development, other phrases have recently appeared. For example, the working paper on reforming the environmental protection law states:

> Nowadays, the common development model of developing countries depends on the exploitation of natural resources, use cheap labor, cause environmental pollution and lack sustainability. The problem of climate change creates the chance to change development thinking, looking for a model and a way to develop free from carbon, sustainable[6] (Working Group for the Reform of the Law for Environmental Protection 2020).

The meaning of sustainability remains the same, but it is detached from the downsides of development, namely exploitation of the environment. A paradigm shift shows that the concept of Sustainable Development is not static but remains under discussion within state organizations. In the process of reforming the law, the board of advisors suggested dismissing the term Sustainable Development from the law text completely. At the time of writing, the new law has not yet come into force, so it is unclear if the suggestion was taken into consideration.

The economic understanding of Sustainable Development is reflected in the human-environment relation, namely the environment with humans and the environment without humans. In the former, humans handle, use, exploit and tame the environment. In the latter, humans are not visible but are implicitly present as actors in relation to nature, and the term "environment" (*môi trường*) is often changed to "nature" (*thiên nhiên*). The concepts in the government paper, therefore, differ from those in this research.

The narrative of humans with nature is defined as the utilization of natural resources by humans. Humans are seen as stewards of nature, and the environment as a support system for development: "Environment is a system of material elements, natural and artificial, that influence the preservation and development of humans and living creatures" (National Assembly of the Socialist Republic of Vietnam 2014).[7] What is significant here is the

---

6. *"Hiện nay, mô hình phát triển thông thường của các nước đang phát triển là dựa trên khai thác tài nguyên thiên nhiên, tận dụng lao động giá rẻ, gây ô nhiễm môi trường dẫn đến phát triển thiếu bền vững. Vấn đề biến đổi khí hậu tạo cơ hội để chúng ta thay đổi tư duy phát triển, tìm ra mô hình và phương thức phát triển theo hướng phát thải các-bon thấp, bền vững."*

7. *"Môi trường là hệ thống các yếu tố vật chất tự nhiên và nhân tạo có tác động đối với sự tồn tại và phát triển của con người và sinh vật."*

division: human, nature and ecosystems do not belong within each other but are separated into different actors, coming together to form the environment. Distinction is made between natural and artificial, which further explains the difference between nature and the environment—the latter includes things man-made.

A quote from Project VIE (1989) underlines the division of humans and the environment and their paradoxical relation: "The population of Vietnam paradoxically constitutes the country's most valuable renewable resource as well as the greatest threat to its environment." Thus, humans are integrated into the economic sphere, but at the same time they are seen as the cause of environmental destruction.

While humans are separated from the environment, they are still thought of together when discussing solutions for environmental problems. However, this does not happen when it comes to the term nature. Nature is mostly used on issues concerning environmental protection, biodiversity strategies and protected areas. Humans are excluded and not seen as part of nature. As McElwee (2016) notes, the term nature is also used by the government to maintain its power over the people by displacing them or changing their livelihoods. For example, after some forest areas in Central Vietnam are declared as national parks, humans are not allowed to reside in the protected areas and not allowed to collect food and other products from the forests for their livelihoods. The relocation of the former park residence allows the state to integrate them into the market and include them into the administration of nearby local communes (McElwee 2016). The state's denial of the political agenda beyond nature conservation makes it possible to mute critics on the assumed motives.

Besides its emphasis on economics, Sustainable Development also places importance on social politics, which is crucial for socioeconomic development. The social aspect deals with the locality of the discourse by including local problems and their definition by the CPV. It focuses on disaster risk management and the adaptation and mitigation of "natural disasters" and their consequences on livelihoods, including migration, poverty and the vulnerable groups.

Health is also a reoccurring theme in environmental laws and policies. An improved health care system is seen as a result of climate change adaptation and as a win-win situation for environmental protection: "Combine the control and reduction of pollution, improve the environment, preserve nature and biodiversity; put people's health as priority" (National Assembly of the Socialist Republic of Vietnam 2014).[8] This emphasis on the public health care system creates friction with neoliberal patterns and underlines the strong role of the state in ensuring social benefits for its people. It, therefore, draws from the party's legitimization under socialist ideology as the caretaker of the people and addresses urgent and visible problems in the current societal debate.

Over the last decades, less focus has been placed on poverty reduction and eradication of hunger. Another issue that is no longer mentioned is overpopulation. In 1989, the central state policy was reflected in the National Plan for Environment and Sustainable Development (Project VIE 1989): the goal was to "achieve a population level and distribution that is in balance with natural sustainable productivity and standard of living" through "implementation of effective population stabilization programs". In 2004, there were still mentions of how controlling the birth rates was important for environmental politics. But in the past ten years, there has been no further mention of this issue. Instead, as Vietnam has managed to lift a large part of its population out of poverty, the discourse of green lifestyle and green consumerism has become the centre of the social pillar of Sustainable Development in Vietnam.

This trend mirrors the shift in power exertion over the decades. While the state used to rule overwhelmingly through its apparatus, it has gradually shifted to doing so through the citizens' conducts. To adapt to the market economy, citizens needed to be not only good socialists, but also good capitalists to build a prosperous nation. This means that while the broad vision has remained the same—a socialist state with prosperity for everyone but no poverty—what this wealth looks like and how it can be obtained have changed. The responsibility for obtaining wealth has been partly allocated to the citizens via their self-optimization and self-entrepreneurship. The capitalist mode of production and consumption has been "greened", and thereby individual consumption choices become crucial.

Another policy discourse that has gained importance over the years is security. While energy security has already been included in policies in the early 2000s, the concept has widened to include food security in 2011 (Office of the Prime Minister of the Socialist Republic of Vietnam 2011) and national security in 2013 ("ensure national defence and security";[9]

---

8. *"kết hợp kiểm soát, khắc phục ô nhiễm, cải thiện môi trường, bảo tồn thiên nhiên và da dạng sinh học; lấy bảo vệ sức khoẻ nhân dân làm mục tiêu hàng đầu".*

9. *"bảo đảm quốc phòng và an ninh của địa phương".*

Central Committee of the Communist Party of Vietnam 2013). In Vietnam's Nationally Determined Contributions (NDCs), security is a key term in 2015 (Socialist Republic of Vietnam 2015). The use of the security frame mirrors an international shift in the security-environment nexus in which climate change and other environmental crises have been introduced as non-traditional security threats over the years. As a country that has met with aggression from major global powers throughout its history and is on constant security alert in the South China Sea and the Mekong region, Vietnam has built its foreign policy on multilateralism. It actively seeks the support of allied forces and continues to do so in non-traditional security threats. This demonstrates another interconnection between different scales and discourses from different policy fields.

International cooperation matters for security and beyond. Strong emphasis is placed on the different roles played by the international community in Vietnam's environmental policies and their influence. As early as 1991, the Vietnamese government made it evident its need for cooperation with external actors when the foreign ministry appealed for investment and support from abroad (Chair of the Cabinet of the Socialist Republic of Vietnam 1991). In 1994, the government stressed that national environmental lawmaking had to be in line with internationally signed agreements and stated that international laws took precedence over national laws in cases of conflict (Socialist Republic of Vietnam 1994). Official development assistance (ODA) continues to be central in Vietnam's foreign cooperation. In its Intended Nationally Determined Contributions (INDCs) statement, the government states that the national budget covers only one-third of finances needed to implement necessary climate change adaptation and mitigation measures (Socialist Republic of Vietnam 2015). Vietnam relies on foreign funding and rejects paying for environmental-related costs and expertise (Schirmbeck 2017). It wants the UN to publish a working definition of Green Economy (Socialist Republic of Vietnam 2012, p. 19). In return for receiving financial and technical support, Vietnam shares its own experiences with the international community.

> Vietnam looks forward to the continued multi-sided cooperation and assistance from the international community, particularly from international organizations and developed countries, with a view to overcoming challenges and moving forward toward a green economy, further improvement of the institutional set-up, hence continued sustainable development of the country. With this report, Vietnam wishes to share with the international community its own

experience in sustainable development implementation, while looking forward to the international community's continued cooperation and support, especially financial and technical support, with a view to continuing its efforts for sustainable development given the new crises in the 21st century (Socialist Republic of Vietnam 2012, p. 19).

The reliance on international financial resources and expertise and the desire for autonomy at the same time is another example of the complex relations beyond the national scales described in Chapter 3.

An important aspect of Vietnam's involvement in international environmental policy is its role in implementing the REDD+ (Reducing Emissions from Deforestation and Forest Degradation in Developing Countries) programme. Various papers have stated the REDD+ programme as a key activity in Vietnam's climate change mitigation efforts. However, the programme has been criticized for being too economically focused and for perpetuating neocolonial power structure (DiGregorio et al. 2015). Sikor (2013) argues that government interventions on ecosystem services have immediate effects on social justice and governance, and they also impacted human-environment relations. The weakness of the programme is that problems of social justice remain inadequately addressed. At the same time, the wide definition of forest limits the ecological aspect of sustainable forestry. These flaws lead to inaccurate planning, monitoring and evaluation of the programme, making it difficult to ascertain if its goal of preventing deforestation is achieved (McElwee 2012). Unintended consequences, like replacing pristine forest with plantations, are not taken into consideration. Despite its shortcomings, the national and international governance actors continue to focus on the programme because it, nevertheless, reaffirms ecological modernization and consolidate power and strengthen the economic nexus of Sustainable Development.

What I have discussed so far are the discourses adapted for the global scale and the commitment to international frameworks, as well as the desire for international cooperation. The discourses in the papers result from incorporating national and local concepts and needs. For example, the Directive to Implement the National Plan for Environment and Sustainable Development aims to "satisfy the basic material, spiritual and cultural needs of all the people of Vietnam ... through the wise management of natural resources" (Chair of the Cabinet of the Socialist Republic of Vietnam 1991). Another paper on The Law on Environmental Protection states that "environmental protection must accord with natural, cultural and historical laws and characteristics and suit the level of socioeconomic development of the country" (The President of the Socialist Republic of Vietnam 2005). Yet, the Government of the Socialist Republic of Vietnam (2017) stresses harmony among all aspects of Sustainable Development within the local context for the Vietnamese people. Historic and cultural references are rarely mentioned, but they nationalize the concept of Sustainable Development and are used for policy justifications.

There are only subtle differences in the languages used in and the intended readership of government papers. Overall, the terms and discourses used are translated either from English into Vietnamese (e.g., Sustainable Development, modernization and Green Growth) or from Vietnamese to English (mostly state-centred language such as "propaganda").

There are two exceptions to word-by-word translation. The first is from *phát triển kinh tế xã hội* to socioeconomic development. The Vietnamese version emphasizes the economic development aspect, and this is reflected in the national policy. This term is well influenced by international flows and policies but is much less integrated into international politics and discourses than Sustainable Development. The English translation, however, does not completely encompass the width and depth of the Vietnamese term.

The second exception is the non-translation of the terms CSO and NGO. The Vietnamese government hardly uses the term CSO, and therefore it is surprising that this term found its way into the Report on Implementation of Sustainable Development. A possible explanation could be that the report was presented at the RIO+20 conference and therefore had a wider target audience than the usual government report (Socialist Republic of Vietnam 2012). In Vietnamese language publications, the official terms used are socio-political organizations, mass organizations and social-career organizations, thereby bringing all actors under the state policy. Indeed, the Vietnamese government has the intended audience in mind and uses language strategically to fit a certain context, thereby contributing and co-shaping discourses on different levels.

The process of localizing the discourse goes beyond translation and is used to clarify and reinforce national power structures. For example, the government stated in Agenda 21:

> The people of Vietnam are not restrained in their consumption of natural resources for food and other purposes by the dictates of religious, moral or traditional taboos. The great respect which previous generations showed for the balance of natural forces and all living things including the forest spirits,

has been lost or forgotten as a result of the terrible social upheavals that took place during thirty years of war and because of the sheer demands of the growing population for more immediate production (The Prime Minister of the Socialist Republic of Vietnam 2004).

The quote exemplifies the government's argument that environmental degradation is caused by individuals and a lack of environmental awareness among the citizens, which must be countered by educational measures and centralized state power. Following the party's slogan—"people know, people discuss, people implement and people supervise"—the responsibilities to protect the environment lie in the government and every individual (Ministry of Natural Resources and Environment 2008). To do this, the CPV exerts power by decentralizing and maintaining control at the same time.

Overall, the Report on Implementation of Sustainable Development summarizes the role of state institutions in balancing economics, societal developments and the environment within the country and beyond, focusing particularly on the Mekong River and ASEAN:

> The institutional framework that Vietnam pursues must ensure harmony among the economic, social and environmental fields of sustainable development, constituted by the national institution and in keeping with international practices. It must also be implemented in vast territorial regions (economic regions and river basins) with a view toward development for human beings in the future (Socialist Republic of Vietnam 2012, p. 17).

Although both China and Vietnam have similar environmental concepts, there is one major difference between them. China has created its own environmental state narrative using the concept of "ecological civilization", or in short, "eco-civilization" (shengtai wenming 生态文明). While Vietnam uses the term "eco" on the local level, it has not used it as an alternative to Sustainable Development in the national policy.

The "ecological civilization" narrative was first introduced in 2007 and then endorsed by President Xi Jinping in 2013 as a framework for China's environmental law-making and execution (Hansen, Li, and Svarverud 2018). It is a short-term policy framework and at the same time a long-term vision for China's development and even the global future (Hansen, Li, and Svarverud 2018). "Ecological civilization" is already implemented across different levels in China, forming everyday environmental rule in accordance with Chinese philosophy as a counternarrative to Western discourses. It is "a socialist-ecological future with Chinese characteristics" (Jiang 2013), although it is still based on socio-technical understandings of environment, combining neoliberal economic strategies with redistribution efforts (Hansen, Li, and Svarverud 2018). It is, furthermore, used to underline the state's legitimation to rule and take responsibility beyond its own nation-state to solve environmental crises.

Vietnam has not established a similar counternarrative to reintroduce emic philosophy and replace global narratives with national ones. Instead, it takes the Sustainable Development narrative and attaches its own agenda and meanings to it to enforce the state legitimacy as shown throughout this chapter. It is, therefore, committed to the global scale and international treaties, reaffirming its desire for a multifaceted foreign policy in which aligning with different centres of power remains key. This enables Vietnam, for example, to obtain financial support for climate-related investments through international regimes. It enhances Vietnam's negotiating power: when different offers for renewable energy investments are on the table, Vietnam can negotiate the terms and conditions of the deals better. In the changing global world order, Vietnam does not seek dominance, but diversified allyship.

NGOs and the Sustainable Development Narrative

For this part of the narrative analysis, I interviewed NGOs and read their publications to examine how they perceive environmental issues and the Sustainable Development narrative. Most of the NGOs handed me their publications when I met them for interviews. In cases where they did not have any print publications available, interviewees pointed me to their websites to download their latest publications. I then applied a grounded coding approach to select outlying and noteworthy patterns and frictions in and among narratives that became apparent during the process. Grounded coding stems from the idea of grounded theory and is a tool to evaluate qualitative data by building categories for the analysis from the data given (Glaser and Strauss 2017).

The field of environmental NGOs is, of course, broad and diverse, so are environmental topics themselves, which cover many aspects of the relationship between humans and their surroundings. The NGOs selected for this book work on climate change, disaster risk reduction, pollution, sustainable agriculture and food security, wildlife conservation, biodiversity protection, water, renewable energy and green lifestyle. Their actors and narratives differ.

One distinction among organizations working on environmental topics lies in their approach to the environmental problem: either explicit or implicit. Taking an explicit approach means that an organization sees itself first and foremost as an organization that works towards seeking changes and impacts on the environment (e.g., climate change mitigation, natural resources conservation and forest restoration). This approach does not consider long-term visions but focus on an actual goal that can be achieved by the organization. On the other hand, taking an implicit approach means that the organization works towards its goal through activities that impacted the environment. This could mean supporting women's rights by empowering women to manage the forest in their community or decreasing the vulnerability of poor communities by enhancing climate change adaptation projects. It is important to note that explicit and implicit approaches are two ends of a spectrum. There are organizations that use both approaches in different projects or where lines are blurred between the two ends.

The NGOs selected for this research use both explicit and implicit approaches. Organizations that use the explicit approach are WWF, Greenhub, IUCN, ENV and C&E. Greenhub aims to "minimize adverse impacts for ecosystems and biodiversity" through their programme on plastic litter in coastal areas (Greenhub n.d.); ENV seeks to end wildlife trade and protect endangered animals (ENV n.a.); C&E's mission is to "protect the environment, conserve natural resources and promote sustainable development" (C&E 2015b). These narratives separate humans from nature and perceive humans as active agents and nature as a passive recipient of these actions. Scientific approaches are used as solutions to environmental problems, seeing the responsibility of environmental crises on multiple levels.

The explicit approach is mostly used on the national and international scales. Local projects are more concerned about social issues (e.g., poverty reduction and gender equality), but the location of implementation is defined by the problems and internationally approved discourses.

Organizations that mostly follow the implicit approach are ActionAid, CARE, SRD and Oxfam. For example, SRD works on sustainable agriculture and livelihoods, climate change adaptation and disaster risk reduction, forest law enforcement, gender equality, child protection and support people with disabilities. It uses agriculture and food security issues to solve both environmental and social issues: for example, its climate change adaption projects improve income and reduce poverty, and at the same time enable forest areas to recover. ActionAid promotes sustainable livelihoods so that more people (especially women and the disabled) can participate in land and forest management, and this also reduces deforestation.

CENDI is an example of an organization that uses both explicit and implicit approaches. It promotes environmental protection of conservation areas and recognizes primary forest ("*rừng nguyên sinh*") as the home for endangered species (CENDI 2019). At the same time, its narratives intersect with alternative narratives (e.g., spiritual ecology and land rights) to focus on the multilevel, mutual interaction of nature and humans. Furthermore, land, or places, is portrayed as an actor in the sociopolitical system where humans interact, and this creates hierarchies and power structure.

In Chapter 3 of this book, NGOs use narratives that relate to their theories of change and their understandings of the environment and nature. When Vietnam first allowed the establishment of international NGOs and social organizations, the focus was on poverty reduction. The field has since widened to allow state and international actors to intervene in local contexts (Dewan 2021). During this process, international organizations saw the need for local NGOs to increase their capability. The report from UNDP and MPI (1999, p. 19) recommends that INGOs "help Vietnam NGOs with capacity building. Governmental, non-governmental and intergovernmental organizations provide information and technical support, share experiences, and train staff for Vietnamese NGOs so that they will be able to contribute effectively to the process of sustainable development."

Vietnam's development, which has been closely connected to its economic growth and industrialization, came along with its integration into global economy and politics. This resulted in more buildings being built, heavier traffic on the roads, increase in population and greater digitalization. Almost all interviewees have seen both positive and negative aspects of these developments. While they appreciate the reduction of poverty and accumulation of wealth across society, they are worried about environmental destruction and increasing capitalist production and consumption. The environment and economy have been pitted against each other, and only through the Sustainable Development discourse has this trade-off been resolved using the ecological modernization concept.

While the interviewees acknowledged the changes happening in their surroundings, they differ in their opinions on how far these changes have affected the human-nature relationship. On the one hand, all interviewees argued that the people are less aware of nature and no longer perceive its intrinsic value. On the other hand, most interviewees said that the people are more aware of nature because they are more educated now. These two paradoxical trends became clearer when I asked the interviewees for their own connection to nature. City dwellers reported making efforts to reconnect to nature, for example, by taking weekend trips to the green spaces in Hanoi. Others reported that online purchases and increased consumerism lead to waste generation, which reduces their environmental consciousness. It is not surprising that both trends exist in parallel since both behaviours are reflected in the official discourse.

When I asked the interviewees about the biggest environmental concerns currently in Vietnam, they mentioned similar issues: pollution and waste crisis. Climate change was only mentioned occasionally. When I asked why climate change was (not) mentioned, the respondents said that although climate change is an issue of concern, it is not a "Vietnamese" issue. This means that they perceive international actors as responsible for climate change mitigation since Vietnam is unable to work on the climate crisis independently. Therefore, interviewees see local and national issues as the most pressing environmental problems in Vietnam. No difference could be seen in the responses between INGO and VNGO staffs.

Pollution has increasingly impacted the lives of people living in Vietnam over the last decades. These impacts mostly refer to health problems, such as respiratory diseases and cancer, and less to the effects on the environment. Despite the adverse effects on human health, some interviewees felt that the effects on human-nature relationships are inevitable. One respondent explained that children in rural areas used to learn swimming at a young age, but nowadays they hardly do so as the lakes and rivers are too polluted for swimming. Similarly, due to the high levels of air pollution, children in urban areas are hardly allowed to play at the parks, and therefore lack experiences with nature. Thus, the pollution problem is framed as a social concern, without assigning blame to any specific actor.

The waste crisis was more likely to be brought up by international staff, but when it was mentioned by Vietnamese staff, they were more likely to frame it as a plastic issue and not an overall waste problem. The topic of plastic waste has gained traction in both NGO field and public discourse. This took place alongside a global debate about plastic in recent years, which led to different initiatives around the world, including the ban on the use of single-use plastic. It is interesting to examine the global plastic discourse across different sociopolitical contexts. In Vietnam, plastic waste on beaches and the seemingly apolitical nature of the topic, as well as international funding by donor organizations, helped to raise awareness in all parts of society. It seems easy to tackle the plastic waste issue within existing institutions; moreover, the issue matches the government's *"xanh sạch đẹp"* narrative. Since plastic waste can easily destroy clean and beautiful spaces, therefore, anti-plastic campaigns align well with governmental narratives and public discourses. In constrast, the international dimension of exporting waste from the European Union to Southeast Asian countries is hardly mentioned by the NGO staff. Although plastic waste has a global dimension, it is still framed as a national issue.

There are different causes for environmental problems and hence different solutions are offered. Three groups of actors—individual citizens, government institutions and systemic reasons—are deemed to be responsible for environmental problems. Vietnamese NGO staff highlighted the role of the individual citizens. The rise of the "convenient" consumer lifestyle adds to the waste problem as some people buy things that they do not need. Hence, citizens (*người dân*) should change their habits, for example, by recycling waste and disposing them properly. The interviewees observed that more people bring their own shopping bags and use less plastic straws nowadays, therefore they are becoming more aware of environmental problems. The Vietnamese term *ý thức*, translated as awareness or consciousness in English, implies educating the people, and it is connected to the idea of civilization (Harms 2016, p. 89). This means that the urban-middle class sees others, especially those living in rural areas, and not themselves as contributing to the environmental situation. One interviewee wished that the people could consider how their lifestyle choices could impact the future generations and be proactive to make changes and not wait for the government to take the lead. Focusing on the individual behaviour can be interpreted both as a Vietnamese tradition and as an understanding of the individual's role in a socio-economy.

Government institutions are the second group of actors responsible for environmental problems. The laws enforced by these institutions are insufficient to promote the use of renewable energies or ban wildlife trade, and they are not strongly enforced. However, they are important because they prevent chaos and promote equity, and refrain people from bad behaviour. In the long term, the people expect environmental rule to play a bigger role in Vietnam. The people's rights and their freedom show a paradox in contemporary Vietnam (Harms 2016). During *Đổi Mới*, liberalization led to more freedom granted to some actors but less freedom for others. Nevertheless, the NGOs continued to advocate for more freedom to adjust power relations in the state. Their networks and theory of change determine the rights that they advocate. Granting rights to the people means giving them the freedom to form networks so that they can tackle waste problem in their commune on their own. It also enables citizens or communes to sue companies over pollution issues, which would prevent companies from polluting the environment and penalize them for emissions.

Interviewees were divided on their views regarding discursive ideas and systemic problems behind environmental concerns. On the one hand, interviewees praised Vietnam's economic development. On the other hand, they felt that the new consumption trend that resulted from greater wealth was a key problem. The Sustainable Development narrative—correlating a higher living standard with a form of consumption that is sustainable and in line with Vietnam's development narrative—is used to overcome this contradiction.

The perceived friction between development and the environment is cited as an underlying problem. Due to economic growth, consumers' demand for goods and energy are on the rise. Economic growth is seen as more important than environmental concerns. The interviewees rarely mentioned economic actors, and only two of them who worked for NGOs promoting alternative environmental narratives were critical of the capitalist economic system.

The main challenges faced by the NGOs when working with governmental institutions are hierarchies, lack of transparency and corruption. Due to their low salaries, most government officials rely on additional income either by taking up extra jobs (e.g., consultancies which are often paid by donor agencies) or by demanding compensation for their existing duties. Some NGOs refuse to be involved in corruption while others see no way around it. The government's hierarchical structure results in decisions made at the top level, even if the work is done on a lower level. Together with the high level of bureaucracy, this increases the processing time needed for obtaining approval. The international NGO staff said that the longer wait might be a reason why there are only a few national NGOs in Vietnam. Hence, hurdles for working in the Vietnamese institutionalized framework are high. Consequently, this strengthens a top-down approach, which translates across actors and influences how NGOs approach their projects in the communities.

"So I think that [my organization] is the place that gives back values to the community in this way" (*Thế thì mình thấy [tổ chức của mình] là cái nơi mà nó mang lại những gía trị cho cộng đồng như vậy*), said one NGO staff member during our interview. The interviewees' values and views on nature and the environment were more clearly revealed during the interviews than in publications, workshops and conferences. The causal settings of the interviews seemed to allow space for the interviewees to connect their emotions and values to the issue of nature and the environment.

Although the interviews allow space for personalization, Sustainable Development continues to be framed in terms of ecological modernization. Similar as in the publications, words connected to Sustainable Development and environmental awareness—"*khai thác tài nguyên*" (extract natural resources), "*kinh tế xanh*" (green economy) and "green lifestyle"—were repeatedly mentioned throughout the interviews. One interviewee made clear, however, that Sustainable Development is not the same as Green Economy. Sustainable Development has a much broader meaning, and the government and its development partners understand this term differently. They see it as a process rather than a definition. Indeed, when examining projects and processes that are framed as Sustainable Development, it becomes apparent how differently Sustainable Development is implemented.

Unlike in government publications and statements, NGOs do not use strong economic language. Government discourse, as well as those of donor organizations, connects Sustainable Development to green growth, markets and economic stakeholders, while the NGOs place less emphasis on the economy-environment nexus. NGOs' economic goals are not macroeconomic, but part of social strategies and are not connected to GDP growth. NGOs frequently use the following terms in their publications: payments for environmental services, REDD+ programmes, low-carbon markets and natural resources. These terms are used to frame and contextualize local projects into the national narrative.

All NGOs use the word "sustainable", although not necessarily paired with "development". "Sustainable development" is mostly used to frame a project to a larger context and to formulate the NGO's missions. For example, C&E's missions are to "protect the environment, conserve natural resources, and promote sustainable development" (C&E 2015b). Its visions also point to how sustainable can be used without development: it "actively participate in solutions to environmental problems, conservation of natural resources, and maintenance of sustainable life" (C&E 2015b). Its goals are to "build sustainable livelihoods and conserve local natural resources; advocate environmental and development policies, and monitor government policies and projects" (C&E 2015b). Other NGOs use "sustainable" in agriculture, consumption and energy. Sustainable livelihoods emphasize the social aspects at the local level and will be discussed in the following chapter. Overall, the word "sustainable" is used to indicate a social-ecological connection and indicates a win-win situation instead of a trade-off between different concerns. Furthermore, all organizations use "sustainability" and "sustainable development" to connect their work to the global sustainable development goals (SDGs). Sustainable Development is a narrative tool to frame processes across scales and is universal to incorporate different contexts into the process.

Exceptions to using the Sustainable Development narrative are found in the social spectrum. For example, the just transition narrative between GreenID and Friedrich-Ebert-Stiftung (FES) underlines social justice; the co-benefits narrative used by GreenID and the CCWG is concerned over issues of well-being, health and employment. These alternative narratives are discussed below.

Additionally, socioeconomic development is used as a framework for interventions, although to a lesser extent than the Sustainable Development framework. The government uses socioeconomic development narrative to position the organization or project within its national strategy. The same is true for the Green Growth strategy, which is used by both Vietnamese and international NGOs to justify their projects. For example, GreenID uses Green Growth strategy in addition to SDGs and the Paris Agreement, and Greenhub uses the socioeconomic development narrative as a basis for their work in coastal conservation.

Sustainable Development is not seen to exclude other narratives. Instead, it is combined with other terms that in some contexts are perceived as conflictive and antithetical to its concept. This happens frequently in the Vietnamese context. For example, Sustainable Development is mentioned together with climate justice, co-benefits or socioeconomic transformation. When organizations collaborate with each other, their narratives are consolidated. For example, CCWG uses the climate justice and Sustainable Development narratives, and when it collaborates with an international donor organization that uses the socioeconomic transformation concept, this is also included in the narratives. Thus, the choice of narratives depends on which actors are involved in the process. Therefore, the Sustainable Development narrative offers space for organizations to adjust themselves and their work according to the network activated.

Like the government, NGOs define Sustainable Development according to their purposes. They understand the concept as economic regarding natural resources (*tài nguyên thiên nhiên*) and as human agency regarding environmental protection (*bảo vệ môi trường*). C&E (2018b, p. 23) exemplifies the environment as a service provider: "Nature provides for us necessary power sources", "eco-system and ... eco-system services".[10] GreenTrees (n.d.) calls the environment a "place of dwelling and provider of information for humans".[11] Service provider is related to the concept of ecological modernization.

Some publications mentioned undesired outcomes of Vietnam's economic growth, such as depletion of natural resources, deforestation and pollution. However, this does not result in a criticism of development or economic growth per se. Instead, the NGOs argued for a proper implementation of sustainable strategies and the need for Green Growth initiatives. Alternative discourses, such as the human-nature connection, are almost exclusively used in projects involving ethnic minorities. For example, C&E uses the Sustainable Development narrative but adopts an alternative approach regarding forest management for ethnic minority communities. In this chapter's case study, the NGO also applies the Sustainable Development narrative on their projects and introduces ethnic minority communities and their knowledge as alternatives to common development practices.

Intrinsic values, such as health, form supportive arguments for socio-environmental change. They are used in such a way that they do not pose a threat to the system. For example, GreenID (2018b, p. 24) quotes health as a co-benefit for energy transition and as a reason to reduce air pollution ("green air, clean lungs"). According to CECR (2018), water pollution needs to be addressed to reduce the impacts on health and the economy.

NGOs' Sustainable Development narrative has a strong social connection, just like in the government discourse. Sustainable Development is often linked to vulnerability and poverty reduction. Vulnerable groups are defined as women, ethnic minorities and communities living in areas most affected by climate change or other ecological impacts. It remains to be seen how this

---

10. *"thiên nhiên cung cấp cho chúng ta các nguồn lực cần thiết"*, *"hệ sinh thái ... và dịch vụ sinh thái"*.
11. *"nơi lưu trú và cung cấp thông tin cho con người"*.

emphasis on vulnerability empowers the selected groups (Sivaramakrishnan and Agrawal 2003; Li 2007). Poverty reduction is also strongly linked to climate change adaptation. Dewan (2020; 2021) describes how climate change adaptation has replaced poverty reduction as a buzzword in today's global political context. In Vietnam, NGOs practise climate change adaptation and mitigation on the local, national and international levels. Climate change, vulnerability, resilience and disaster risk reduction are often interlinked. Alternative discourses like climate justice, in contrast, are almost completely absent. Development brokers often have a conventional understanding of development as a form of progress measured against Western values of modernization. This understanding allows actors to work towards a goal that they perceive as worth achieving.

This raises the question of who are responsible for environmental problems. The answers obtained from interviews differ from those found in the publications. While most of the interviewees saw getting the people to adopt environmentally friendly behaviour as a solution to environmental problems, the publications focused on achieving change through policy (GreenID 2018c; CARE 2019; SRD 2018a; ActionAid Vietnam 2021). Different views were expressed due to differences in purpose, context and target audience. While a personal conversation allows room for debates, enables an individual to express their viewpoints and is only self-representative, publications undergo an editing process, are tailored for the public and represent the organization.

Generally, the term "green" has similar meaning to "Sustainable Development", and it is used by several different organizations. It is found in the names of GreenID and Greenhub. These two organizations use this term as a corporate identity for their projects and to position themselves in the national narrative as the green branding is easily recognizable.

Green Trees uses the same tactic. Interestingly, it is the most critical of the government among the interviewees. Members of Green Trees have been persecuted, have their passports withdrawn and have been put under pressure. However, from the interviews, they are not as critical of the government as they are perceived to be. Green Trees's goal is not to completely change the system, but to work within the environmental law. Its narrative can hardly be understood as critical: Green Trees positions itself within the government's definition of a green and sustainable development.

NGOs use science and technology to understand nature and the environment and to provide solutions. They frame their projects and goals in a technical and quantitative way. For example, SRD (2018b, p. 16) states that "Approximately 180 households in 8 boards belonging to 2 wards of the project VM059 in Son La have implemented the CSR rice farming technique (smart rice cultivation for climate change adaptation) with an area of 200-500m2"[12] and "Reduction of 2-3 times spraying pesticides and no use of herbicides in implementation of the model".[13] The NGOs use the same approach to provide solutions. For example, Greenhub's monitoring and assessment programme states that "The monitoring results will be analysed. We will use this to share and recommend for general application to collect national datasets for marine waste monitoring in Vietnam which can be used as inputs for governmental policy recommendation" (Greenhub n.d., p. 1).

The Sustainable Development narrative has manifested both Vietnam's global position and the CPV's legitimacy. NGOs in Vietnam have adjusted to power relations in the authoritarian context as well as to norms on the global scale by reproducing narratives about the environment. Defining environmental problems and finding possible solutions hardly differ between actors, but they are nevertheless diversified because of the different narratives used.

Alternative Environmental Narratives in Contemporary Vietnam

Besides the Sustainable Development narrative, a few actors have attempted to establish alternative narratives but with limited success. This section discusses these alternative narratives, how they differ from the Sustainable Development narrative, which actors are involved and why. Failing to establish alternative discourses helps to understand why the Sustainable Development narrative remains strong. The ambiguity of the Sustainable Development narrative allows room for localization, which is not seen in other narratives. Other narratives' critique of the political systems prevents them from being adopted in Vietnam and poses risks for actors who actively pursue them. Thus, these alternative discourses are not as widespread as Sustainable Development.

---

12. *"Khoảng 180 hộ gia đình ở 8 bản thuộc 2 xã vùng dự án VM059 tại Sơn La đã áp dụng cấy lúa theo CSR (canh tác lúa thông minh thích ứng với BĐKH) với diện tích cấy theo CSR từ 200-500m²"*,

13. *"Giảm được 2-3 lần phun hóa chất trừ sâu bệnh và không dùng hóa chất trừ cỏ khi áp dụng mô hình 17"* (SRD 2018b, p. 16).

Justice Narratives

Outside of Vietnam, the justice paradigm has become an important narrative that connects grassroots movements and institutionalized organizations across the world. It has formed narratives around environmental justice, among them climate justice and just transition. The environmental justice movement, as well as the just transition movement, originated from the United States and has been adopted by different actors globally since the 1990s and 1980s respectively. In both cases, marginalized groups were at the forefront of the movements—Black and persons of colour in environmental justice and workers in just transition (Jafry, Helwig and Mikulewicz 2019, p. 2; Schlosberg 2007, p. 80). Climate change was seen using a justice or rights lens to examine the causes and consequences (Bell 2020, p. 3). Later, it led to more general environmental questions and connected them to social concerns (Bell 2020, p. 13). Climate justice can be perceived as a practice, a concept of normativity, or as an object of formal inquiry (Tamoudi and Reder 2019, p. 65; Boran 2019, p. 27)—or even as a narrative that clusters actor-networks in relation to multiple centres of power.

The definitions and dimensions of justice vary among approaches. The "empirical approach" includes three dimensions of justice: distribution, participation and recognition (Schlosberg 2004). Other authors have emphasized the legal aspects to justice (Natarajan 2021). A representative from an international NGO stressed that their conceptual framework of justice includes representative, procedural and recognition justice, which must be understood through a feminist and postcolonial framework. How justice is achieved and which social groups are at the centre of its definition vary from case to case. For example, the postcolonial lens has criticized the Western hegemonic concept of nature-human relations in the environmental movement and the lack of representation from the Global South. Northern NGOs used to focus mainly on conservation and preservation of environment in a way that excluded humans from their definition of environment and nature (Cha 2019, p. 212). Human interests and environmental interests were seen as separate. The environmental justice movement then changed the narrative to understand social issues as being interrelated with environmental problems.

The focus on justice and the connection to human agency has led to criticism. Opponents of environmental justice understood this concept as yet another separation of humans from nature and possibly leads to the neglect of environmental concerns (Derman 2019, p. 420). On the other hand, proponents of this concept argue that the idea of environmental justice challenges and questions the meaning of environment and opens up space for different understandings of the environment and justice (Schlosberg 2013, p. 2). For the Global South, "the global justice perspective is a welcome relief from the formerly post-materialist dominated agendas of Northern environmentalists" (Wong 2012, p. 46). Wong explains how NGOs like WWF have shifted their agenda and analysis from exclusively conservation to also include social issues, thanks to the environmental justice movements.

Centring on the justice concept, several other movements and their discourses have been developed. For example, energy justice is a subfield of the environmental justice movement. Energy justice is the "application of rights (both social and environmental) at each component part of the energy system" (McCauley and Heffron 2018, p. 2 cit. in Fuenfgeld 2019, p. 224). It focuses on ethics in the energy field that is characterized by "power, distribution of and access to resources" (Healy and Barry 2017, p. 452).

Another network within the justice movements is the labour movement, which shapes the narrative of just transition. According to Cha (2019, p. 214),

> just transition can be understood as the conceptual framework with which the labor movement captures the complexities of the transition towards a low-carbon and climate-resilient economy, highlighting public policy needs and aiming to maximize benefits and minimize hardships for workers and their communities in this transformation.

The labour movement came about because of a struggle by unions and workers' associations in carbon-intensive and extractive industries (Rosemberg 2010; Healy and Barry 2017, p. 454). Seeing workers and unions as part of an environmental justice network opens new networks, although conflicts have arisen whereby the workers accuse environmental activists of hijacking their movement. Other definitions of the just transition narrative are much broader. Swilling (2019, pp. 3–4) refers to just transition as a "commitment to eradicating poverty in our lifetime without destroying the planet's natural system" and does not focus on the role of the labour movement, but instead include movements from the Global South that are committed to social justice.

The just transition paradigm has focused on the workers' rights during times of economic changes. It has been adapted widely by policymakers in the last few years: for example, the International Labour Organisation has established guidelines for a just transition; the international climate negotiations (Conference of the Parties (COP)) have adopted a just transition work programme; and the G7 countries are creating Just Energy Transition Partnerships (JETPs) with Vietnam. The term—just transition—has entered the political mainstream and its meaning has been contested by different actors. The Vietnamese state has defined just transition using three characteristics: access to affordable energy; historical climate justice; and reskilling and social programmes for workers in the fossil fuel sector. The just transition discourse has been successfully adapted because it does not question systemic power dynamics. It dissolves the connection between carbon and growth, and energy and development, and recognizes the advantages of a low-carbon economy (Newell and Mulvaney 2013, p. 132).

Progressive INGOs contest the narrow focus of just transition and demand for the inclusion of other aspects such as distributional justice and gender justice into the concept.

A few VNGOs, such as the CCWG and GreenID, use justice narratives in their publications. Projects that used justice narratives are all funded by and implemented together with international organizations. One such organization is FES, which is the affiliated foundation of the Social Democratic Party of Germany (SPD). SPD perceives itself as working with labour unions and workers' organizations and has therefore adopted the just transition narrative. It also shares FES's vision of establishing an environmentally and socially just world without renouncing capitalism. These organizations use the justice narrative to place themselves in an international debate. Their publications and local projects use either the Sustainable Development discourse or other narratives that were introduced by donors. Hence, it can be assumed that the justice narrative is used to fulfil donors' expectations. Donors and INGOs seek to influence Vietnamese local discourses by introducing narratives used globally. The justice approach is sensitive in a repressive political context because it is used to criticize power relations and injustice. International cooperation projects open limited space for this approach as the justice issue is framed globally and not within the Vietnamese context.

Although the climate justice approach connects development with environment, it also links human rights (Derman 2019, p. 420). The human rights discourse is sensitive for the Vietnamese government, although it is in line with the government's approach of framing environment using socioeconomic development. As McGregor et al. (2018, p. 497) state, "climate justice cannot be considered to be 'post-political'" and herein lies the main difference to Sustainable Development. Sustainable Development does not specify responsibilities while climate justice assigns specific solutions to specific problems (Healy and Barry 2017, p. 454). This contrasts with the Vietnamese government's framing of ecological crises as management problems with a modernist understanding. Sustainable Development also does not discuss possible contradictions between economic development and climate actions, but rather links them together. On the other hand, the justice approaches open the friction for discussion.

Transformation Narratives

Another group of narratives in NGO discourses are centred around transformation. These narratives share an epistemological history, although they have since developed in different directions, thanks to diverse actors and schools of thought (Brand and Wissen 2017). Swilling (2019) divides the transition and transformation narratives into two different schools. The first school states that structural changes are possible within the current system by using socio-technical improvements and with individuals playing an important role (Bell 2020, p. 11). The just transition narrative belongs to the first school. The second school, where most proponents of the social-ecological transformation belong, sees the need for profound changes in all areas of life (culture, economy, politics) and an ultimate renunciation of capitalism and current development concepts (Bell 2020, p. 5).

The second school bases its arguments and understanding of transformation on Karl Polanyi's *The Great Transformation* (1978). Polanyi explains how economic models have taken over social concerns, and this is extended to the ecological sphere too. Today, proponents of the socioecological transformation argue that we are facing another turning point (Wolf 2008, 32ff). The socioecological transformation is an ongoing, reflexive negotiation process that questions power relations and seeks to transform society, economy and nature (Brangsch 2008, p. 9; Dellheim and Krause 2008, pp. 19–20). Its goal is to build a free and united society amid a stable biosphere. Transformation serves not only to determine norms but also to analyse problems and find solutions (Brand 2012, p. 52).

Brand (2012, pp. 28–29) says that the concept of Sustainable Development is as contradictory as Green Economy because they bring opposing interests together into one narrative. Comparatively, socioecological transformation is less contradicting as it does not focus on some areas, such as economic growth. Using the socioecological transformation approach, different analytical schools seek to address problems and propose solutions. Among them are academic debates on degrowth, practice theory and political ecology (Brand and Wissen 2017). This approach has the advantage of leaving space for interpretation, but it is criticized for being diverse since places are still understood via economic models and natural resources (Degenhardt 2016, p. 30).

The Rosa Luxemburg Foundation (RLS), a German political foundation that is active in Vietnam, shifted its narrative from Sustainable Development to socioecological transformation as it uses an alternative approach to development (Degenhardt 2016, p. 5). Socioecological transformation is a counternarrative to Sustainable Development and its implications of green growth and green consumerism; instead, it proposes an alternative vision for society and economy, including the redistribution of power.

The concept of socioecological transformation is adopted by organizations cooperating with RLS. GreenID used the concept in a project funded by RLS, but it also adopted the co-benefits narrative from another donor organization, showing flexibility in its narratives. RLS tries to promote the concept in several other cooperative projects. Some partners (e.g., C&E) do not adopt the concept and continue to use the Sustainable Development narrative. When cooperating with academic institutions, NGOs actively discuss narratives and concepts, especially in their application to the Vietnamese context. NGOs, such as GreenID or C&E, do not perceive themselves as setting discourses but rather act according to their own understandings of the terminology. The research institute that cooperates with RLS, however, had problems with understanding socioecological transformation and even changed the term to socioeconomic transformation at some points in its publication. This shows that the concept is new in the Vietnamese context and there is a strong emphasis on economy. Socioecological transformation is seen as complementary, and not antagonistic, to Sustainable Development.

Socioecological transformation is hardly used because it has mostly been an academic concept and has not been translated beyond academic actors. Vietnam should embrace socioecological transformation since it is a socialist country. Yet, since 1986 Vietnam has followed a capitalist market economy approach with a socialist direction—for which Sustainable Development seems more suitable. Applying socioeconomic transformation can lead to criticism of the Vietnamese state's legitimacy.

Co-benefits

The concept of co-benefits introduces intrinsic values into the discussion on climate policy. This concept is not contradictory to the existing ones but is an addition to widen our understanding of human-nature relations. Co-benefits have appeared in the climate and energy discourse and are connected to the other narratives. They are used to evaluate cost-benefit, thus putting other concerns, such as health, well-being and social injustice, on the back burner. They are also used to influence the political decision-making process towards more environmentally and socially friendly regulations (Ürge-Vorsatz et al. 2014, p. 550). The idea of co-benefits is derived from co-impacts, co-costs, trade-offs, externalities and spillover effects (Ürge-Vorsatz et al. 2014, p. 551)—these concepts do not explicitly touch upon climate issues but are implicitly related to climate policies. For example, renewable energy decreases greenhouse gas emissions and hence contributes to better health; it also empowers communities by decentralizing electricity production.

According to Ürge-Vorsatz et al. (2014, p. 554), the advantage of co-benefits is their adaptability to local contexts and applicability across scales. They look at different groups and their potential co-benefits, therefore making an analysis more thorough. They develop strategies for actions and empowerment along the people's needs and exert pressure on decision-makers. At the same time, they can be adapted to different climate change scenarios by taking trade-offs into account (Workman et al. 2018, p. 674). On an international level, co-benefits can also include national security interest, and traditional and non-traditional security threats. Co-benefits are a narrative and a tool of analysis. They offer space for negotiation for a better world, both within and outside the current political economy.

So far, co-benefits are mainly used in Vietnam to reduce the use of fossil fuels and greenhouse gas emissions (Smith 2013, p. 4). For example, REDD+ projects, which aim to reduce the amount of greenhouse gases in the atmosphere, use co-benefits or "non-carbon benefits". The co-benefits narrative connects low-carbon development to other traditional development discourse, including poverty reduction. Although reducing greenhouse gas emissions and poverty reduction are crucial, they are not the goals of co-benefits (Brown 2019, p. 267). In addition, climate change mitigation itself can be a co-benefit (or climate benefit) when projects reduce greenhouse gas emissions even though they have different priorities (Jacob et al. 2013, p. 6).

Brown (2019) argues that there is no comprehensive and coherent approach to monitoring and evaluating co-benefits. While emission reductions are monitored, the perceived co-benefits are not systematically verified (Brown 2019, p. 267). Mayrhofer and Gupta (2016) add that co-benefits are similar to the business-as-usual approach and do not call for a deeply rooted transformation, which is why they can be applied in the Vietnamese authoritarian context.

Among the Vietnamese NGOs, GreenID uses the co-benefits approach in their cooperation with the Institute for Advanced Sustainability Studies (IASS). The CCWG has also applied the co-benefits rhetoric in its cooperation with IASS. In a GreenID publication with FES under the title of "Just Transition", the report picks up co-benefits as an argumentative and a strategic tool:

> All relevant state agencies in the areas of socio-economic, energy, climate change, green growth, and education should develop policies for the post 2020 period that consider the co-benefits brought by the energy transition. These opportunities for economic growth and social justice must be integrated into the national policies and sectoral strategic plans for sustainable and equitable development (GreenID 2019, p. 16).

Co-benefits are a narrative but have been adapted as an argumentative tool. It has not become a competitor to Sustainable Development. The values promoted under co-benefits could challenge the ecological modernization paradigm, but they can also be addressed under the Green Economy nexus.

Sustainable Livelihoods

Sustainable livelihoods is a concept closely linked to development and poverty reduction. It emerged in the 1990s as a counter concept used to perceive rural development (Levine 2014, p. 1). It analyses how people live in their respective sociocultural context (Scoones 2009; Levine 2014). Sustainable livelihoods is therefore a framework, a concept, a tool, a set of principles and a way of understanding used by organizations in the development sector to connect rural development, poverty reduction and environmental management (Scoones 2009, p. 3; Ashley and Carney 1999, p. 8). This concept is mainly concerned with how livelihood strategies relate to their resources and wider contexts (Scoones 2009, p. 3). Thus, one needs to consider institutional processes and individual behaviour change that include elements of adequacy, security, well-being, capability, resilience and natural resources. All these elements are used to identify ways to improve the lives of the poor, which translate into concrete development activities (Serrat 2017, p. 21; Ashley and Carney 1999, p. 1).

Connected to achieving the Millennium Development Goals (MDGs), sustainable livelihoods is used to achieve the SDGs. The linkage between socioeconomic and environmental concerns within the sustainable livelihood narrative was established in 1992 at the UN conference: donor organizations tend to use sustainable livelihoods as an operational tool, thus underlining the vulnerability and resilience approach in climate change adaptation (Brockelsby and Fisher 2003, p. 186; Cannon, Twigg, and Rowell 2003, pp. 3–4). Embedded in the sustainable livelihood approach, natural hazards are the main obstacles to poverty reduction.

Vulnerability is an analysis of risk exposure, both short-term and long-term. It predicts the future and is highly political in definition. Development actors have used it almost interchangeably with poverty and shifted vulnerability (Dewan 2021, p. 18). The meanings of vulnerability (*sự dễ bị tổn thương*) and resilience (*khả năng phục hồi*) have become an important nexus for NGOs, VNGOs and INGOs alike. While the Vietnamese translation of vulnerability frames the agency of people as vulnerable; resilience is translated as a skill that everyone could possibly acquire. Climate change adaption is a social lens for framing climate change consequences, while climate change mitigation is perceived using a techno-scientific framing. Thus, NGOs in Vietnam have taken on resilience and vulnerability in their social goals and visions.

There have been various criticisms of the sustainable livelihood approach and its underlying narrative. First, it is criticized for excluding some aspects of the people's lives and for connecting power with discrimination. For example, in the initial framework, gender relations are excluded (Levine 2014, p. 2). Proponents responded that the framework can be adapted for every context. Second, it is deemed too complicated for workers to implement in their projects. Proponents argued that the actual situation is not easy to understand. Third, the sustainable livelihood approach is criticized for emphasizing the economy, and this diminishes other aspects of life (e.g., it subordinates social relations to economic ends). Fourth, its critique of vulnerability and resilience perpetuates potential weaknesses and does not recognize actions, reasons and consequences. It neglects strengths and agency. Proponents supported introducing changes on an individual level and proposed solutions to uplift the people through economic means.

NGOs rooted in poverty reduction and "classic" rural development use the sustainable livelihood narrative in their publications. For example, ActionAid Vietnam uses it almost exclusively, even though they also use the Sustainable Development narrative.

The sustainable livelihood approach focuses specifically on a target population and approaches environmental issues from a social perspective. It is compatible with the Sustainable Development framework. Sustainable livelihoods is concerned about the relationship between poverty and the environment, whereas Sustainable Development is concerned about how the people can live in a way without compromising the ability of future generations to meet their own needs. Scoones (2009, p. 5) points out that unlike sustainable livelihoods, Sustainable Development has to find "uneasy compromises". But the basic assumptions of what sustainability means and which strategies to use are not contradictory. Vietnam's policymakers recognize poverty as an issue and eradicating poverty has brought legitimacy to the CPV over the last decades. Therefore, the Vietnamese government wants to uplift approximately six per cent of the population out of poverty (World Bank 2021) and to do it in harmony with environmental goals. This sustainable livelihood narrative works well for a diversity of actors: NGOs which are concerned with poverty eradication and environmental concerns; government parties that grant permits for organizations; and the international development community from which this concept is derived.

Local and National Narratives in Connection to the Global

All these narratives that are used in the publications analysed for this research entered Vietnam from an international scale, which consists of networks of international environmental policy, international development work, NGOs and donors. The global scale produces counternarratives, but they are hardly present in Vietnam. These counternarratives are nationalized through donor relations as they are used only in the respective projects with donor organizations. For example, the CCWG uses the justice approach, which forms the bridge between the international and national levels by consulting decision-makers on NDCs and the climate negotiations.

What about other national and local narratives that exist within or beyond the Sustainable Development paradigm? Vietnam's national discursive frame is socioeconomic development, which comprises Sustainable Development and other environmental narratives that are adjusted to the Vietnamese context. Sustainable Development even functions as a tool to bridge international and local discourses.

As mentioned, local discourses of the environment are narratives related to ethnic minorities. Different framing and alternative local narratives are seen through the process of othering. For example, a clear distinction is made between Kinh environmental behaviour and their socioecological structures and those of ethnic minorities. This othering process makes separate discourses for those who are positioned outside of the networks of power and those who are within the networks. The ethnic minorities have different narratives from the Kinh because they are governed differently. The process of othering also puts ethnic minorities at a distance from the state; they do not fall under the same governing discourse that NGOs position themselves in. The ethnic minorities perceive themselves as outside the state, which contrasts with the assimilation strategies used elsewhere (Scott 2009). Local concepts are hardly applied on the national level.

In his research on China and Taiwan, Weller (2006, p. 169) found that there are fewer local discourses than international ones. He concluded that the imbalance is because the local discourses have a "far more limited sphere of influence" and "they have not generated many serious attempts to present themselves as alternate sets of ideas". Instead, the local discourses are practised in their respective contexts but are not applied on another scale. They remain local and therefore do not compete as an "alternative environmentalism" with other discourses (Weller 2006, p. 170). Despite having a limited discursive power, local discourses can keep diversity alive and adjust to changing times and locations. In addition, they do not stay separated and isolated from international narratives. The different discourses meet, overlap and influence each other. Although certain terms and phrases are seldom picked up and introduced into the national discourse, Sustainable Development may be practised in ways that do not openly contest the actual understanding of the term, but nevertheless takes shape in different meanings according to different localities.

Occasionally, we see alternative narratives that relate with Vietnamese cultural history. They include the earlier examples of the spirit world when advocating to maintain urban trees and the role of nature in Vietnamese belief systems. The Green Growth strategy brings modern aspirations as compatible and complementary with Vietnamese traditions through the Sustainable Development narrative:

> The rich and beautiful traditional lifestyle is combined with civilized and modern means to create comfortable, high quality and traditionally rooted living standards for people and society of a modern Viet Nam. Implementing rapid and sustainable urbanization while maintaining the living in harmony with nature in rural areas and establishing sustainable consumption behaviours within the context of global integration (Green Growth Strategy 2012, p. 3).

This illustrates how morality is created with nationalization. The government decentralizes economic and environmental efforts but maintains control over them. This top-down approach of environmental authoritarianism remains strong. It is, however, not uncontested and not singular. The following third vignette illustrates the contestation of environmental authoritarianism when practising the Sustainable Development narrative.

Case Study 3: Challenging the Narrative through Practice: Doing Sustainable Development Differently in Central Vietnam

The hills were in bright green, with a forest cover that stretched out as far as I could see from our car. It was a beautiful scenery; the afternoon sun beamed, warming the landscape and making the rivers in between the hills glitter. Having left air-polluted Hanoi a few hours ago, I felt that I could finally breathe again. I seemed to have arrived at a nature place. But Chi (all names in this case study have been changed to protect anonymity), the founder of several NGOs, an activist, an intellectual and a powerhouse in her sixties, made me realize how wrong I was in reading the landscape. She pointed to the trees around us and explained: "This is all monoculture." We were standing in the middle of acacia plantations that were cultivated for paper production. A forest is not simply a forest. That was the first important message on my field trip to Central Vietnam.

The NGO was founded in the early 2000s, but its history goes back further and is closely linked to Chi, its founder. Chi became involved in environmental issues when she attended university in Vietnam and got into trouble with her research. Nevertheless, she continued her engagement and advocacy for land rights and was against environmental destruction. The government decided in the early 2000s that Chi was "a dangerous woman" (her own words) and ceased all her work. Chi had to lie low for a while, until the legal framework and the lawmakers offered her space for action once again. Today, two of her organizations remain active; Chi is still involved in them although there are staff who take care of the daily operation.

When I met Chi, she was immediately eager to help me with my research. I was not the first PhD student to show up at her office, but her dedication to the cause of her work and to academia opened her door for me. On a rainy autumn day, she took me to a popular tourist destination, which was one of her project sites in Central Vietnam. For over twenty years, Chi has been working with several communes in this area, supporting them in acquiring land rights and establishing "agroecology" on their land.

According to Chi, the relationship between society and natural resources has not recovered since the American War. The war caused great destruction of resources, especially forests. Following that, the implementation of *Đổi Mới* and Vietnam's integration into the global development community in the 1990s led to new concepts and efforts to restore Vietnam's damaged forests. But this reforestation effort was carried out top-down and during the development efforts, natural resources were once again destroyed. Donor organizations supported the destructive practices due to corruption and their lack of knowledge and understanding. The villagers were not consulted as they were portrayed as backward, and any consultation was done mostly through government agencies. Ethnic minorities' beliefs and practices were labelled as "superstitious" and "backward" and pitched against the "civilized" model of development.

Today, the donor and international cooperation landscape has diversified. Both Chi's NGOs receive funding from different organizations. Still, Chi remains critical of other organizations. When I met her, she was trying to settle a dispute with a non-Vietnamese public investor. The investor's state-run forest protection project had encroached onto the territory of one of the communities that Chi has been working with for years. The project seemed to neglect the local community's customs and ways of using forest land. But after a series of negotiations, an agreement was reached. Both sides praised their projects as "sustainable", but the meanings of what sustainable mean differed. For the investor, it means increasing the forest-covered area according to the definition of the Vietnamese government, which considers both natural and planted forests as the same.[14] For Chi, it means giving space to local communities to practise their worship of nature.

---

14. Circular No. 34/2009/TT-BNNPTNT on the criteria for determining and classifying forests.

Over the years, Chi's approach for change has remained consistent, but her position in networks and relations to power has evolved. She is actively working on developing her networks closer to positions of power. In the commune that we visited together, the head of the communal party chapter and the lead of the Fatherland Front were farmers whom she worked with on land rights and agroecology over twenty years ago. In other areas of the country, the contacts she has supported over the years have made their ways into the ministries.

This does not mean that her work is welcomed everywhere. Some provinces are still reluctant to grant permission for field site visits for her visitors, and she is cautious about what she says and which communication channels she uses to pass on sensitive information. While I noticed her confidence, she seemed to be constantly on alert as she has witnessed the enforcement of authoritarian state power on land rights.

Chi is not only conscious of her communication channels, but also mindful about the content and words of her communication. She consciously switches narratives when working with different organizations. In one context, she spoke about agroecology and worship of nature; in another context, she framed the same content under Sustainable Development; and again in another context, she used socioecological transformation. The ideas and project goals remained the same, but she needed to fit into the organization's narrative to obtain funding and cooperation. For example, she told me some farmers were tricked into planting cassava for a one-time pay-off by a large company. She described the problem as an issue of land rights, Marxism or spirituality according to different settings. When she spoke to a potential donor, she classified it as "food sovereignty" so that it would fit into the donor's programme. Eventually, she was successful in securing their cooperation.

The same is true for the NGO publications. These publications cover a wide range of narratives with different target groups but share a common goal: granting power to the people and suggesting alternative development concepts. Under this goal, the NGOs combine the Sustainable Development narrative, which focuses on growth targets and natural resource management, with agroecology, which focuses on diversified land use, land rights and the position of spirituality in the human-nature relationship. These parallel and sometimes contradictory narratives lead to frictions: they communicate different values, but the NGOs use this coexistence to build connections and networks with the local ethnic minority communities and the government institutions. The NGOs choose narratives to suit the target group so as to achieve their goal. In this case, the goal is to obtain communal land rights and restore the communal land with agroecology instead of plantations. Consequently, the NGOs use the Sustainable Development narrative towards state agencies, although they reject a modern understanding of ecology. They recognize that they need allies in the state structure to navigate the authoritarian context.

When I visited project sites in Central Vietnam together with Chi and two other team members from the NGO, I noticed that the NGO used a direct approach to change power relations between humans, institutions and nature. The NGO has a main head office in Hanoi and a second head office in Central Vietnam. The second head office is seen as equally important as the main head office, and it is near the main project site; hence, it serves as both a geographical and an ideological centre of the NGO's networks. The NGO wants to build a farm for practising agricultural techniques and for farmers from different provinces in Vietnam and other Southeast Asian countries to network. The project site is home to forests, fields, a river and a diversity of plant species. There is also a seed bank to preserve seeds of plants from different parts of Southeast Asia. The site serves as a place to train farmers and for government officials and farmers to meet. Hence, it is a place for both practical and theoretical education, and for the key farmers' programme.

The key farmers' programme has been a basis of the NGO's work way before the existence of the organization. In all the communities where the NGO or its predecessors have been active in, a few individuals have been identified as the key farmers. These key farmers are from the communities and function as multipliers. They attend training courses, where they meet key farmers from other areas, exchange experiences on farming techniques and ways of dealing with the local authorities, and even travel across the country to carry out this informational exchange. The programme has helped to instil a sense of project ownership among the communities, enabling them to continue for more than twenty years in some places. It has also helped the NGO to carry on their operations, as some key farmers have come to hold party or government positions, granting continued opportunities for project permits.

A sense of ownership struck when I visited the community. The farmers' identities seemed to be deeply connected to their struggle for land rights and alternative development. The NGO translated the farmers' identities to actors outside of the community as personal life stories. Here I briefly recount the stories of Trang and Thanh.

Trang is a young mother who has spent all her life in the community. Her grandparents worked with Chi when she first started these projects twenty years ago. They own a piece of forest with a pepper plantation where Trang used to play as a young girl. She grew up learning from her grandparents on how to take care of the "forest garden" and about the plants and their uses. Trang was happy to explain to me the uses of the different plants, and she also pointed out the birds and their relation to the ecosystem. However, after completing her studies, Trang decided not to be a farmer. She worked as a schoolteacher for a few years and later got married. But she continued to find herself in a dilemma because she was not able to make enough money from her job as a teacher. Despite this, she refused to take bribes from her students as corruption was against her values.

Eventually, she decided to quit her job as a teacher and took over the pepper forest from her grandparents. Her parents-in-law and her husband disagreed with her decision because they perceived a farmer as having a lower social status. Trang got divorced, which is rare in Vietnam (Tran 2021). With the help of her grandparents' connection to the NGO, Trang was able to secure financial assistance to expand her farming business and start a permaculture garden, growing vegetables and fruits for her own use and for selling at the local markets. A break with gender narratives goes hand in hand with a break in narratives of farming practices. While she is not consciously seeking to impact systems through her actions, Trang has nevertheless assumed a role model figure. Farming techniques that are perceived as backward secured her a livelihood, which has allowed her to break with social norms and gender stereotypes.

Thanh's story also started before his generation. His father was one of the first key farmers, but he is now the head of the party chapter in the commune. Similar to Trang, Thanh initially decided against farming and also against staying in the commune. Like many other young men from the area, he left for jobs in urban centres. Thanh even went as far as Singapore to work as a construction worker. When the COVID-19 pandemic broke out, Thanh was left without an income as construction sites were temporarily shut, leaving him with no choice but to return to the commune. Upon his return, he became involved in farming because he saw it as a more stable and safer way to make a living. One year after his return, Thanh has become one of the most active commune members in the NGO project, giving it his own spin by setting up social media channels for the project. On Facebook, he regularly posts self-made videos (he even bought a drone for making these videos) and pictures from the gardens. He wants the viewers to see farming as something positive, and not as something backward. Thanh rethinks his aspirations and seeks to reframe agriculture. He thereby challenges hierarchies and stereotypes that are widespread in Vietnamese society.

"*Người quê*" (countryside people) are generally perceived as backward and not modern (Nguyen 2014). Rural areas are understood to be the roots of families; they have become places of the past from which young people migrate to cities to build their modern future. Thanh's return to the commune and his attempt at telling a different narrative use the idea of a future that is connected to ecological sustainability. Members of the community are conscious about the general aspirations of modernity in Vietnamese society, and they are consciously seeking change through building new narratives and practices. However, not all practices associated with modernity are rejected. Instead, practices evolve to make sense for actors in their context. There is no rejection or adaption of practices simply because they belong to a certain narrative. Through this approach, practices co-shape narratives from the ground.

During my visit to the commune, Chi took me to visit all the key farmers and their model project gardens. The NGO has been able to determine which land belongs to whom, assigning eight hectares to every family and enabling individual households to decide on their agricultural practices. Having gained access to their land, the key farmers turned the vegetations around. Instead of growing only acacia, they applied different farming techniques to see what worked best with the local soil. When we visited Sao in his model garden, I received a crash course from him and Chi on the goals and needs for this model of agriculture. In agroecology, nothing is wasted and everything—every plant, every craft, every material from humans—is dependent on each other:

> ... the unique activity of agroecology. We came here and everything is acacia but how raise awareness to the farmers in this area. Why? Everyone wanting money. Everyone make contract with the acacia company. Every 5 years they have to give acacia to the company. So we have to select carefully knowledgeable farmers to supervise them and ok the steps to reform acacia to the diversity of the species, to the ancient trees, herbal medicine plant so you can harvesting short seasonality to earn short time money. Or bananas. And now they are very instinct about banana rings because after 6,7 months

bananas grow up very quick and to sell to the market to get money. And that's why you see Mr. [Sao's] garden. He starts to replace acacia with different things and trees and incense and herbs and different types. And now very interesting that he attracts the young who are philosophers who have a business idea and now they come back because they have 4 hectares, the parents hand over to them and now they learn from Mr. [Sao] to do step by step.[15]

Chi and her NGO acknowledged the different needs and sought to satisfy them in the project. They recognized the wishes of farmers to earn money quickly, provide for their families and be part of the wider development around them. Again, the realities on the ground intersect with societal narratives. Modernity meets environmental degradation and deprivation of land rights. The community is therefore open towards changes that promise economic stability (guaranteed through short-term income generating crops like bananas), ecological harmony (long-term agroecology farming techniques) and empowerment through reclaiming their rights. The mixed cultivation serves the health of the soil and the overall ecosystem, connecting social and ecological concerns and communicating limits and boundaries on what is possible in this model.

Together, we visited the forest pepper plot owned by Trang's grandparents. The commune members who came along shared their understanding of nature with me. For example, in distinguishing between different kinds of jackfruits, Chi and Trang discussed the differences between forest jackfruit and garden jackfruit. Chi depicted the forest jackfruit as "a natural jackfruit" (*mít tự nhiên*), but Trang replied that she also grew this kind of jackfruit in her garden, so it can be neither totally forest nor totally natural. The human intervention is not supposed to be present when something is labelled as "natural", causing confusion about whether the one that grows in the forest can be called "natural" or not. Further discussion led to a debate on which kind of jackfruit can be sold on the market and for which price, expanding the definition to economic value. Reflecting on what is nature, Chi explained further:

People say nature is toxic, people say nature is healthy, nature is foolish, nature is wise but nature does not say so! That's why people say: "Yes, I live in harmony with nature". That's not possible. He says he lives in

---

15. The original English text has not been edited.

harmony with his wife, that's possible, he lives in harmony with his husband, that's possible, got it? You can't live in harmony with nature. Sorry sir, but you have to rely on nature. Nature is not broken to be created. Stop talking about land creation. Say instead that recovery is the strength of the soil.[16]

Chi touched on the relationship between nature and human and the agencies that define both, rendering human action as an intrusion into what she perceived as nature. She used the idea of a dichotomy between human and nature. Consequently, she argued that development is only possible through an interaction between human and nature, and not human mastery over nature: "Nature is diverse, specific, interactive, adaptive and sustainable. Regarding people, this is sovereignty, land sovereignty, this is a sovereignty over land worship, this is local knowledge, this is a local pepper, and this is the management." Again, she created a political ecology by interweaving power with nature.

Chi pointed at an altar that stood at the foot of a tree and requested the group to take a picture: "In a moment, we should take a picture with the space around the altar, in the middle of the forest please. I really like altars [*laughs*]. Sao, but when I die then I request to rewrite my will to not be put up on the altar." In her definition of sustainability, Chi also included the altar to reflect her ideology:

Is this garden a place for a diverse ecosystem? Is there any particularity? Is there interaction? Is there adaptation? What do these four create? Sustainability. It cleans the eco-system of the commune right? The humanist aspect, does the owner of this garden have a red book yet? He has already. So, does this garden have an understanding of worship to nature yet? Already, the altar is here. Right? Is there any local knowledge yet? Is it within its regional context yet? Already. Is there a managing body yet? See, so five criteria of a [human] family and five criteria of nature are benchmarks for this product.

16. *"Con người nói thiên nhiên độc, thiên nhiên lành, thiên nhiên dại rồi thiên nhiên khôn nhưng thiên nhiên có nói vậy đâu! Cho nên con người nói: "Ừ tôi sống thuận với thiên nhiên." Không được. Ông nói ông sống thuận với vợ thì được, ông sống thuận với chồng thì được, được chưa? Ông không thể nói ông thuận với thiên nhiên được. Xin lỗi ông, ông phải nương tựa vào thiên nhiên, thấy chưa [...] Thiên nhiên có hư đâu mà cải tạo? Cải tạo là cải tạo con người ấy. [...] Đừng nói cải tạo đất nữa. [cười]. Mà nói hoàn phục lại cải sức khỏe của đất."*

So please young friends of the commune, when you introduce a product for sale then say this, this product of us fulfils 10 standards [*laughs*].[17]

Chi underlined the importance of the spiritual and intrinsic elements of nature. She focused on these elements in her projects with ethnic minorities. The fact that this is her first engagement of the non-ethnic minority community using the non-material understanding of nature shows that the discourse exists everywhere. This discourse is embedded and is seemingly intrinsic to the concept of agroecology that Chi preferred to use for her projects. At the same time, she saw the need to frame the garden as marketable and fitting into an economic understanding. This is a first step towards green economy where spiritual nature relations can be used as a selling point.

Linh—a young man who studied philosophy but returned from the city to practise agriculture in the commune—and Sao described how their immediate environment had changed over the last years. They said that the forest where we were located had not changed much, but it had overcome storms and changes due to its "natural balance" (*cân bằng ... tự nhiên luôn*). They found this forest to be a good example for Chi's work. Chi explained that the responsibilities for planting trees, managing watersheds and keeping the forest clean lie in the individuals: "Chi imparts on people that who wants to have a clean environment, a clean breath has to protect the forest, has to plant more green trees, when you eat in the garden you know the design of the system of the water source, how to treat wastewater to make it clean and civilized."[18] This statement reflects an internalizing of "*xanh sạch đẹp*" and the wider government narratives.

Linh saw the need for agroecology. He noticed that a flood occurs whenever it rains, and a drought happens whenever the sun is out. It is difficult for

---

17. "*Vườn này ấy là nó ở trong một hệ sinh thái có đa dạng không? Có đặc thù không? Có tương tác không? Có thích nghi không? Thế nhưng mà bốn cái này tạo ra cái gì? Bền vững. Nó làm sạch cái hệ sinh thái của [xã] này đúng không? Cái hệ nhân văn này, cái vườn này chủ vườn đã có sổ đỏ chưa? Có rồi. Thế vườn này đã có một cái hiểu biết để mà phụng thờ thiên nhiên chưa? Có rồi, bàn thờ này. Đúng không? Có tri thức địa phương chưa? Có giống địa phương chưa? Rồi. Thế có đồng quản trị chưa? Đấy, thế là năm cái của một gia đình và năm cái của thiên nhiên như vậy là 10 định chuẩn của sản phẩm. Thì xin các bạn thanh niên của [xã] khi mà đưa một sản phẩm ra bán thì nói thế này, sản phẩm của chúng tôi là có 10 cái định chuẩn cơ [cười].*"

18. "*có Chi là truyền đạt con người muốn được cái môi trường cái hơi thở nó nó trong sạch thì phải bảo vệ rừng, phải trồng thêm những cây xanh, ăn ở khu vườn thì biết là thiết kế hệ thống của nguồn nước, làm sao là xử lý nước thải như thế nào để cho nó sạch làm cho văn minh.*"

the farmers to handle these changes and at the same time protect the forest while securing investments. Also, the families are uncertain about their share of the land and the exact borders of their gardens. The government blocks their access to land rights by withholding information. They have repeatedly brought up this issue with Chi. While every family has the right to eight to twelve hectares of land, they are unclear where these hectares lie. Also, even if the geographical location is disclosed, the farmers lack the knowledge to transfer the measurements from a map to the actual forest. Besides acacia, planting other crops needs the cooperation of the government and the farmers to define which areas can be replanted. Thus, changing the environment requires a network of power with an understanding of sustainability. According to Chi, this network building can be highly sensitive because different groups serve different interests and have different views on development. Multiple institutions and scales (national, province and communal) are involved in this process, making it much more complex.

Our visit to the commune served a second purpose—scouting for a new project site. Living in a national park with its unique geology, the residents have struggled to have a say in the tourism development in the area. The commune has two caves that the locals used to have regular access. But after the government handed over exclusive rights of the caves to a tourist company, members from the local communities must apply for a permit if they want to visit the caves. Our delegation experienced the inconvenience when we tried to visit the areas. We met a tour group on a camping site next to the cave entrance; the employee vehemently denied us entry to the caves, and instead turned to me (the only white person in the group) and said: "We hope to welcome you to one of our tours in the future." As we left the campground, Chi pointed out the garbage lying around the area, which did not seem to reflect the image of the tour business as a "sustainable tourism" provider.

The group explained that this was why they wanted to start a tourism project that generates zero waste and is owned by the community. Back in the commune, the group discussed ways to realize this project. They were concerned over the possible motives of investors, how to find trustworthy partners and where to obtain finances for the project. They knew that they needed to form coalitions and networks, and power should remain within the commune and not handed over to outside actors. Socialist ideals of the community and the environment need to be protected when entering the market economy.

Together with older key farmers, government allies and the young generation, the group discussed how to find trustful investors to start a sustainable tourism business and how to implement it in the current context. From the perceptive of possible cooperation partners and international donors of the new project, Chi reported that one challenge is how the commune can receive the finances directly as they are not an institutionalized legal entity. She suggested founding a social enterprise. When the group heard the word "enterprise", they immediately disagreed: they did not see themselves as businesspeople but were merely doing their part for the community. Chi explained that a community-based social enterprise is different to a regular business. Institutionalization would be the only way to conform to government regulations and still receive international funding, while keeping ownership of the project within the community. Another option is to set up an agricultural cooperative that could manage land for the project and operate without polluting the environment. Like the private tourism company that was mentioned earlier, cooperation with businesses from Vietnam and abroad must be dealt with carefully as they posed the risk of the commune losing power and the environment destroyed.

Members of the commune do not have issues with the government per se; in fact, some members are the leaders of the municipality. But they have issues with the overall power structures, individuals in the government and past mistakes made by the party. These problems caused the commune and Chi to consider forming coalitions. Corruption is an issue of concern, so is the lack of effort by the party to take its members to task. Interest groups lobbied the government for their own benefits at the expense of the commune, for example on land rights and access to caves.

These two projects illustrate a long-term strategy not only for this commune, but also for the general work in the province. Chi explained, "So I want to develop co-governance by community in forest and limestone mountains, buffer zone of the [national park] and see how they can really develop by relying on Sustainable Development by the commune; not by government, not by company but by national park, by watershed management, and by the people". "Real sustainable development" entails looking at human and non-human actors cooperating with the landscape to change power relations and the overall situation. Sustainable Development is a matter of ownership and rights, which Chi perceived herself acting as the intermediary, translator and catalyst during the process.

The people in the commune were unsure when I asked them for their understanding of Sustainable Development. They said that they did not really understand the concept but continued to associate their organization with the term. Sao explained sustainability as "relying on nature is sustainable" and continued further: "Sustainability is also when if the forest restores then we don't destroy trees to build wooden houses, but see what lives under the canopy of trees... the big wooden trees have medicinal plants for health, to increase human health every time people have diseases to depend on them to live."[19] But he admitted that he did not know how sustainable was connected to development and to sustainable development. Sao saw development as referring to farmers who could not survive on forest produce and suffered hardship. Linh added that he saw Sustainable Development as a way to "meet the needs of the present but also ensure environment and future. So ... this is guaranteed ... for environment, land, air, water, to be clean."[20] Despite their lack of understanding of the Sustainable Development concept, they related it to their own livelihoods, thereby making it practical in their own context. Although their understanding of Sustainable Development is very close to the UN definition, they challenged solving the ecological crisis through the political-economic framework of a neoliberal market economy.

During the group discussion, everyone agreed on the economic dimension of sustainability. To prove that economic sustainability is good, they mentioned that the commune has become self-sufficient in their fruit production and do not need to buy additional produce from outside the village. Trang viewed this independence positively as she and some others used to work in a factory to earn money to buy fruits, and "every time there is a factory, there is pollution". Chi chipped in and said, "See, this is why we need ecological agriculture." Ecological agriculture is a different form of Sustainable Development; it is less concerned about individuals, privatization, decentralization, technology, and more about using systemic reasons to solve ecological crises and placing responsibilities on government actors.

---

19. *"Bền vững nữa là nếu mà rừng khôi phục thì không phá cây gỗ để làm nhà nhưng mà dưới tán cây gỗ... cây gỗ lớn đang có những cây dược liệu để làm cho sức khỏe, tăng thêm sức khỏe của con người mỗi khi con người có các bệnh cũng nương tựa vào đó để sống."*

20. *"đáp ứng được cái nhu cầu hiện tại, nhưng mà phải đảm bảo được cái môi trường, cái tương lai. Nên là... thế là đảm bảo được cho cái... cái nguồn môi trường, nguồn đất này, rồi không khí này, nguồn nước là được đảm bảo sạch sẽ."*

This case study shows that narratives informed the actions of the NGO, which needs to adapt to existing governance that is closely interwoven with the discursive space. When the NGO and the community cooperate with government authorities, the Sustainable Development paradigm is produced. At the same time, different narratives appear among NGO and the community and in publications with donor organizations. There are frictions between the narratives because they clasp different relations in the economic system and therefore the understanding of Sustainable Development. For example, ecological modernization meets alternative models of Marxist analysis, commodification and agroecology. The NGO and the community use these contradictions strategically to achieve their goals and to facilitate change in the local space. Additionally, there are frictions between discourses and practices. Practices are a way to adapt to circumstances that vary from the societal norm, and they challenge narratives too. As a result, looking at both narratives and practices is needed for a complete picture of what Sustainable Development means in Vietnam.

Conclusion: The Role of Narratives in Environmental Rule and State-Society Relations

In this chapter, I have shown that Sustainable Development is universal. It does not allow alternative forms of political economies, but it nevertheless allows space for localization and ownership of the narrative. Alternative concepts lack these traits and are either too specific to one target group or field of action (e.g., sustainable livelihoods) or they question the status-quo of power relations (e.g., socioecological transformation). The Vietnamese state uses its fluid interpretation of Sustainable Development for its environmental governance, meaning that it implements social rules within environmental policies and uses the narrative to create a behavioural norm. This conduct is at the same time embedded in socialist state politics and capitalist market-based economy. Its goal is to create moral citizens who are committed socialists and entrepreneurs, and who chase the common good and their own wealth at the same time. According to Derks and Nguyen (2020), Sustainable Development in this context has taken the moral turn while using technology to further the implementation of ecological modernization.

As a ruling narrative, Sustainable Development consequently forms perceptions of the environment and environmental crises and informs the positioning of NGOs and their actions within ecological modernization. However, the narrative space is not absolute. Alternatives do exist, although in limited capacities. To understand the variety of Sustainable Development, it is essential to take both discourses and practices into consideration.

The relation between alternative narratives and Sustainable Development allows us to draw conclusions as to why alternative narratives are (not) used. What these narratives have in common is that they are not as universal as Sustainable Development. They offer less space for interpretation since they are connected to specific ideas, approaches and political agendas. Some can be integrated into the Sustainable Development paradigm. For example, sustainable livelihoods, co-benefits and just transition do not seek systemic change, and they become parts of the Sustainable Development discourse. Other narratives are consciously placed outside of that narrative and even as competing concepts, opening spaces for discussions. These narratives do not necessarily fit into the repressive Vietnamese context. Whether outside or within the Sustainable Development paradigm, all alternatives are not easily adopted by Vietnamese actors because they are imported ideas and their relevance for the Vietnamese context is limited. Alternative discourses that are proactively applied in Vietnam are those from the local actors.

State-society relations are not a total dichotomy. As I have already pointed out earlier, the state sphere and society sphere are not clearly separated. Still, these categories remain important in understanding how power is structured in Vietnam's environmental rule. The actors' strategic use of narrative points towards their agenda and shows how the state and society interact. The analysis of the Sustainable Development discourse and its alternatives illustrates why there is a need not only to examine the state and society, but also to differentiate actors in those spheres according to the heterogeneity they display and to understand their specific positioning within environmental governance. Environmental rule with its consideration of circumstances beyond the contemporary authoritarian structure provides the analytical space for discussion. Nevertheless, I contend it is important to not neglect the systems. In this book, I used the narratives within governmental mechanisms as a point of departure. The narratives help us to understand power relations, although we cannot disregard the practices in order to fully understand how governance works in Vietnam. Generally, Sustainable Development tells a story of multiple relations between the state and society, between national and international scales and the processes across them, and between frictions where ignorance and contestation appear. Sustainable Development brings everything together and presents it as a consistent whole so that the ruling regime and the CPV can showcase consensus while allowing for contextualization and challenges in Central Vietnam, the Mekong Delta and beyond.

6

Conclusion

In 2011, I was on my motorbike in Hanoi when I witnessed students protesting air pollution and emissions from motorized traffic. Ten years later, in 2021, I was in Central Vietnam and the Mekong Delta and finally able to understand what environmental action means in Vietnam and which narratives and understandings are used. A lot and yet not much has changed since then. Hanoi looks different now, with its new highways, high-rises and SUVs on the streets. My friends now live in condos instead of socialist housing. They continue to strive as socialist citizens and active participants in the capitalist market economy. Power has shifted a little from INGOs to VNGOs, and the political context in Vietnam becomes harder to act in. Environmental action, in general, and climate action, in particular, are more necessary and urgent than ever before, though the action taken is still insufficient. Climate change and the human responses continue to be complex, and other crises have increasingly happened alongside climate change. Mapping the complexity and identifying all the entry points for action remains inevitable; social scientists and researchers have a responsibility alongside natural scientists and their data findings to find ways to mitigate and adapt to climate change. Through my research, I have sought to understand how power structures inform environmental narratives and thereby shape our understanding of the problems and solutions to ecological crises. This research is a clipping from a large landscape of knowledge needed to understand the environmental map of our world with its sociocultural contexts and political-economic implications. It uses the Area Studies approach to first understand a place in its own specificity, without drawing a theoretical blueprint for use beyond the context.

Within this specific context, I have argued that Sustainable Development is a powerful narrative that has become universal. Derived from a long-standing discourse about "development" and the debate around its meaning for different actors, Sustainable Development finds itself in the same tradition and confronted with similar critiques. Deriving from the Brundtland Commission's report in 1987, the term has since spread and has been adapted and filled with different meanings by various actors. It is now central to environmental rule in Vietnam and used by the authoritarian government and NGOs to navigate state-society relations. Sustainable Development is cocreated in this late-socialist setting by a variety of actors in historical and cultural cross-scale relations. Its ecological modernization programme has moralized the development process and transformed socialist citizens into individualized, capitalist self-entrepreneurs.

In Chapter 2, I have outlined the Vietnamese political context and introduced relevant actors for environmental governance. As an Area Studies tradition, I have emphasized on contextualization to minimize the risk of misleading generalizations. Additionally, I have attempted to not only portray nature as a passive background for my analysis, but also emphasize its relevance as an actant. Landscapes have influenced the socialist government even before the establishment of the Socialist Republic.

In Chapter 3 and in accordance with the environmental rule, I have furthermore defined the cultural, historical and cross-scale paths that led to environmental governance in Vietnam. Some of the narratives and understanding of human-nature relations that are embedded in religious and philosophical histories appeared in policy documents as side notes. Different belief systems and schools of philosophy offer different interpretations of the human-nature relationship. A look into historic milestones shows how this understanding has changed, and which international actors impacted it. The French colonial policy changed environmental policies and landscapes through its impacts on agriculture and infrastructure. While Marxism-Leninism is neither "green" or "brown" in its theory, the rule of the Communist Party of Vietnam (CPV) is connected to the industrialization process, which cause environmental destruction. With the renovation politics in the 1980s and the transition towards a capitalist market economy, the ecological modernization programme has gained momentum in environmental policies and decentralization, with the continual use of science and technology.

Moreover, Chapter 3 argues whether NGOs and CSOs exist in Vietnam and how these organizations should be defined. I lay out that these terms can be used, and I define NGOs through three perspectives: structural definition by the state, self-identification by NGO staff members and networks that NGOs positioned themselves in. In the given political context, NGOs are actors in the state's environmental rule. All the NGOs that I have interviewed identified advocacy work as crucial. They seek to change the legal framework by translating the needs of the communities using a rights-based approach, but not transforming the system itself.

NGOs have adapted the Sustainable Development framework and the ecological modernization programme to navigate the authoritarian state rule and donor landscape. Sustainable Development and its related concepts appear throughout NGO publications, proposals and reports. Alternative narratives like environmental justice, just transition and socioecological transformation are found in some projects because they are advocated by international organizations that seek to criticize the political system and introduce alternatives to Sustainable Development. For VNGOs, pushing these narratives can pose a danger if they are too critical of government narratives. Additionally, these narratives may not be connected to the local context, therefore making it hard for VNGOs to localize them. Nevertheless, these concepts are used in certain projects, but are hardly adapted beyond them. In analysing these practices, I emphasize on the negotiation processes between individuals and institutions to understand how state-society relations function.

Chapter 4 shows the need to follow processes across and between the global, national and local scales. Historically, this means paying attention to colonial entanglements that persist until today and inform VNGO and INGO relations. However, there are limitations since international relations can offer advantages too, especially when VNGOs need to make space for themselves in the authoritarian context.

Building on the relations between actors and history of certain narratives, Chapter 5 examines closely the Sustainable Development narrative and shows that it has become universal. The Sustainable Development framework has become a central narrative to environmental state policies, and it connects to social issues and strengthens the state's governance. The Vietnamese government uses Sustainable Development as part of its international integration initiative and perceives its own position and policies within a global effort to prevent ecological crises, while accumulating economic wealth and industrializing the country. These economic implications of Sustainable Development are centred around the importance of growth and a possible reconciliation between liberal economic policies and environmental protection, as well as the perception of the environment as a resource. The state has created norms and values around Sustainable Development that brings together the morality of the socialist citizen with the capitalist, self-optimizing consumer and producer. Sustainable Development, thereby, becomes a part of the socioeconomic development strategy of the Vietnamese government, which uses the framework to legitimize the CPV's policies in the domestic sphere. Sustainable Development becomes a narrative with a strong, social meaning regarding poverty reduction, health and most recently, matters of national security. Government institutions contextualize the global narrative with strategies and policies while maintaining terms inherent in the concept. A successful narrative by the Vietnamese government within the Sustainable Development framework is their slogan "*xanh, sạch, đẹp*". This slogan exemplifies how the narrative is localized and connected to the environment. Beyond government publications, the slogan has consistently surfaced in research and is appropriated and used by different actors. NGOs have referred to it, and interviewees have used it to express their vision of the environment. The idea of modernity inherent to "green, clean, beautiful" has been widely internalized and thereby manifests an understanding of nature that is not wild but is civilized. The pollution of air, water and soil strengthens the "green, clean, beautiful" vision and shapes the understanding of what "developed" means in Vietnam by attributing it to individual behaviour and state action. Other narratives, such as population control, have declined over time, showing the fluidity of the Sustainable Development concept. Sustainable Development with its openness for contextualization and interpretation is defined as universal by Tsing (2005).

While the Sustainable Development narrative has been adapted and is used by all actors that are part of this research, the meaning of the concept still varies. Using discourse analysis, qualitative interviews and participant observation, I have exemplified the similarities and differences between the narrative and its practice. My interviews with NGO representatives have shown that their perception of the ecological problems in Vietnam is similar across actors. Their proposed solutions differ but not radically, as most NGO staff see solutions in better environmental legislature and its actual implementation, and in increasing the people's awareness of environmental problems. Only a few interviewees suggested systemic changes to economy and politics, but not an overall reform of socioeconomic policies. Therefore, the Sustainable Development paradigm follows the government's path dependencies of economic and rights-based policy, although it demands for more freedom and more social security for different groups of actors. Those NGOs that are more critical of the government and work on freedom rights also tend to have different networks from the less critical ones. They are consciously choosing allies for their unofficial networks and are less involved in official ones.

The Sustainable Development narrative is not uncontested, however, as we have seen in Chapter 5. In theory, they adhere to the ruling discourse, but in practice, they opt to do things differently. This shows that it is crucial to look not only at the practice or the narrative of development, but at both.

When we examine the three case studies of NGOs in this book using a comparative lens, we can draw additional conclusions regarding Sustainable Development and the heterogeneity of state-society relations in Vietnam. These three case studies took place in different provinces, targeted different groups and worked on solving a range of overlapping, environmental topics. The NGOs' self-perception of their roles also differs. NGO 3 perceives itself as a partner and a connector. It creates networks that exist between the local communities and between the local and international communities by bringing farmers from different Southeast Asian countries together and by generating funds from INGOs and donor organizations. NGO 3's international network is stronger than its networks within Vietnam and with other NGOs. The people in the province perceive NGO 3 as a partner, an educator, an institution with authority and as a trusted ally. In the national frame, NGO 3 is cautious of its dealing with governance institutions, but it has developed its own network to function within the system but remains critical of it, thus making it an emancipatory group. The organization works strategically with frictions in its narrative and practice in the authoritarian context.

NGO 1 focuses on creating a just society for all, and it uses its international network, finances and reputation not only to enhance Vietnamese voices globally, but also to project marginalized voices within Vietnam. It perceives itself as an advocate, but it is unable to overcome the power structures that it is fighting against, even those within its own organization. Although it is perceived as an ally by the community, it is distant from them. Few people are familiar with its mission and scope because the organization tries to work within the governance structures to be granted permission for work. Like NGO 3, and due to the sensitivity of their work, they are under constant threat of getting their permissions and registrations revoked. Though both organizations share the same strategy of working through other parallel organizations and using personal networks instead of institutional ones, NGO 1 tends to be more careful and includes provincial governments in its project designs. This might be because NGO 1 is an INGO, and it is less critical of hierarchies and power relations.

NGO 2 initiates debates and raises awareness about environmental topics, acting first and foremost as a service provider. It identifies young people as agents of change and therefore organizes spaces for exchange with them: it leads initiatives while leaving room for the university and the students to participate. This agrees with the objective of not inducing systemic changes but to work within given structures to change policy and raise awareness of consumerism and lifestyle choices. NGO 2 is a governance group.

In the NGOs' works, there are frictions among narratives and between narratives and practices. NGO 3 is outspokenly critical of Sustainable Development that focuses on economic growth and large-scale businesses and infrastructure projects; it supports small-scale initiatives and farmers, and alternative agriculture techniques. Nevertheless, it publishes and coedits books on Sustainable Development using the government's understanding of the term: GDP growth and nature as a resource. Although NGO 3 criticizes the governance system, it collaborates with it and even uses it to its advantage. The organization navigates the two sides of the responsive-repressive system strategically, using environmental narratives as a tool to create space for negotiation with the authorities and other actors. At times, NGO 3 uses the Sustainable Development narrative to communicate with international organizations. It switches codes and narratives depending on who it talks to, and likewise, it uses a wide range of environmental narratives in its publications. It accepts contradictions to navigate the systems. When speaking with the community, Chi and other staff of NGO 3 touch on land rights and agroecology, underlining the actual practices in the projects.

NGO 1 mostly speaks about sustainable livelihoods, prioritizing the social approach to Sustainable Development. It uses the environment as a tool for achieving social justice, which is similar to the government's focus on the social perspective of Sustainable Development and the SDGs. Although it criticizes power structures, it is unsuccessful in practising power differently in its projects. Also, it locates itself in NGO networks and works with different donors and INGOs. NGO 1 consults with party officials to induce change within the system.

NGO 2's focus on Sustainable Development closely connects to the necessity for economic growth in Vietnam, but the growth must be "greened" by policies and individual actions and awareness. The question of power and reform is, therefore, a friction in the organization's work. NGO 2 has no interest in changing the system, but it occasionally introduces alternative economic and agricultural systems. All three organizations have found their place within the Vietnamese authoritarian system and the international system of environmental action, although the negotiation for networks and places of actions cause frictions.

All three organizations share the notion that change is to be brought about by two agents—the government through the legal framework and the citizens through their behaviour change—using a rights-based approach in development and an understanding of the individuals as having the responsibility and power to make the system environmentally friendly. The three NGOs feed into making Sustainable Development universal that, despite its boundaries of meaning, still accommodates very different understandings of the environment, nature and theories of change. To fully understand Sustainable Development, it is necessary to look at its discourses and practices. The local situations continue to influence what Sustainable Development means and how it materializes. The combination of a moral, socialist citizen who needs to find their self-entrepreneurial and optimized role as part of a modern society is also reproduced in the environmental sphere, and this renders development projects moral too. Nevertheless, the use of technology in development remains in parallel since this is strongly connected to using science and technology for ecological modernization. Morality and technization, socialist citizen and the neoliberal subject, Sustainable Development and its alternatives are never culture wars, but they are what Tsing (2005) describes as frictions: things that do not quite belong together, but in their interplay sets processes in societies and political systems in motion. They are in the interplay between state-society relations and navigate within authoritarian governance. Nature remains a setting for these processes and is, except for

NGO 3, not foregrounded and not recognized as having an agency. Ultimately, Sustainable Development has failed its ecological pillar while destruction continues. The "will to improve" (Li 2007) remains in the moralization of development projects too. NGOs want to improve their work and the general livelihoods in Vietnam, but to different extents they remain caught in what the international regime allows them to do.

These three case studies, furthermore, show that the scholarly debate on whether discourses or practices of development should be central to understanding power dynamics is not one-sided. Both discourses and practices are crucial to understanding what development means for different actors and how they are used to reaffirm positions of power or challenge them. They influence each other, and it is necessary to scrutinize their mutual relationship to understand the whole picture.

This research is neither a blueprint for understanding Sustainable Development, nor a sufficient anthropology of all actors involved in environmental action in Vietnam. More research needs to be done on the historic understanding of environment in Vietnam and the current initiatives and activities by non-institutionalized actors. My initial political question— whether the current projects are necessary and effective despite the prevalent power relations across different scales—has no easy or single answer. While this book gives a brief overview on these historic developments, it would be worthwhile to examine religious practices as another approach for understanding environmental thoughts and action. Besides looking at religious influences from a philosophical standpoint, the practice of religion could provide valuable data as the communist curb against religion has led to it being grounded as a daily practice rather than a doctrine to follow.

What can be drawn from this research is that power relations limit agencies, but they do not eradicate them. International actors and the Vietnamese state are, nevertheless, in a powerful position in which systemic responsibilities and path dependencies need to be greatly reflected on. More open and democratic dialogues should be held among stakeholders with different narratives to acknowledge and support a diversified and connected effort to meet the environmental crises of our times.

Bibliography

Secondary Literature

Adger, W. Neil, P. Mick Kelly, and Huu Ninh Nguyen, eds. 2001. *Living with Environmental Change: Social Vulnerability, Adaptation and Resilience in Vietnam.* London: Routledge.

Arantes, Virginie. 2023. *China's Green Consensus: Participation, Co-optation, and Legitimation.* London: Routledge.

Ashley, Caroline, and Diana Carney. 1999. *Sustainable Livelihoods: Lessons from Early Experiences.* London: Department for International Development.

Aso, Michitake. 2012. "Profits or People? Rubber Plantations and Everyday Technology in Rural Indochina". *Modern Asian Studies* 46, no. 1: 19–45. https://doi.org/10.1017/S0026749X11000552.

Ayers, Jessica M., and Achala C. Abeysinghe. 2013. "International Aid and Adaptation to Climate Change". In *The Handbook of Global Climate and Environment Policy*, edited by Robert Falkner, pp. 486–506. Oxford: Wiley-Blackwell.

Banse, Gerhard, Gordon L. Nelson, and O. Parodi, eds. 2011. *Sustainable Development - the Cultural Perspective: Concepts - Aspects - Examples.* Gesellschaft, Technik, Umwelt. Berlin: Edition Sigma.

Becker, Egon. 1999. *Sozial-Oekologische Forschung: Rahmenkonzept fuer einen Neuen Foerderschwerpunkt* [Social-Ecological research: Framework concepts for a new major funding domain] 6: ISOE.

Beeson, Mark. 2018. "Coming to Terms with the Authoritarian Alternative: The Implications and Motivations of China's Environmental Policies". *Asia and the Pacific Policy Studies* 5, no. 1: 34–46.

Behrens, Julia. 2014. "Imperialistic Influences on the Vietnamese Discourse of Nature and Environment". Master Thesis, University of Glasgow.

———. 2023. "The Power of Sustainable Development: Environmental Narratives and Power Structures in NGO Work in Vietnam". PhD thesis, Humboldt University of Berlin, Germany.

Bekkevold, Jo Inge, Arve Hansen, and Kristen Nordhaug, eds. 2020. *The Socialist Market Economy in Asia: Development in China, Vietnam and Laos*. Basingstoke: Palgrave Macmillan.

Bell, Karen. 2020. *Working-Class Environmentalism: An Agenda for a Just and Fair Transition to Sustainability*. Basingstoke: Palgrave Macmillan.

Benedikter, Simon. 2014. "The Vietnamese Hydrocracy and the Mekong Delta: Water Resources Development from State Socialism to Bureaucratic Capitalism". PhD dissertation, University of Bonn.

———. 2016. "Bureaucratisation and the State Revisited: Critical Reflections on Administrative Reforms in Post-renovation Vietnam". *International Journal of Asia-Pacific Studies* 12: 1–40.

Beresford, Melanie, and Lyn Fraser. 1992. "Political Economy of the Environment in Vietnam". *Journal of Contemporary Asia* 22, no. 1: 3–19.

Bernstein, Steven. 2005. "Globalization and the Requirements of 'Good' Environmental Governance". *Perspectives on Global Development and Technology* 4, no. 3–4: 645–79.

———. 2013. "Global Environmental Norms". In *The Handbook of Global Climate and Environment Policy*, edited by Robert Falkner, pp. 127–45. Oxford: Wiley-Blackwell.

Biggs, David A. 2010. *Quagmire: Nation-Building and Nature in the Mekong Delta*. Seattle: University of Washington Press.

Boomgaard, Peter. 2007. *A World of Water: Rain, Rivers and Seas in Southeast Asian Histories*. Leiden: KITLV Press.

Boran, Idil. 2019. "On Inquiry into Climate Justice". In *Routledge Handbook of Climate Justice*, edited by Tahseen Jafry, pp. 26–41. London: Routledge.

Brand, Ulrich. 2012. "Transition Und Transformation" [Transition and Transformation]. In *Transformationen des Kapitalismus und darueber hinaus: Beitraege Zur Ersten Transformationskonferenz* [*Transformations of capitalism and beyond: Contributions to the first transformation conference*], edited by Michael Brie, pp. 49–70. Berlin: Rosa-Luxemburg-Stiftung.

Brand, Ulrich, and Markus Wissen. 2017. "Social-Ecological Transformation". *International Encyclopedia of Geography*: 1–9.

Brangsch, Lutz. 2008. "Statt eines Vorworts: Sozialökologischer Umbau als neue Stufe von Vergesellschaftung" [Instead of a preface: Socio-ecological restructuring as new step towards public ownership]. In *Für eine neue Alternative: Herausforderungen einer sozialökologischen Transformation* [For a New Alternative: Challenges of a Socio-ecological Transformation], edited by Judith Dellheim and Guenter Krause, pp. 9–14. Berlin: Karl Dietz.

Brie, Michael, ed. 2012. *Transformationen des Kapitalismus und darueber hinaus: Beitraege Zur Ersten Transformationskonferenz* [*Transformations of capitalism and beyond: Contributions to the first transformation conference*]. Berlin: Rosa-Luxemburg-Stiftung.

Brockelsby, Mary Ann, and Eleanor Fisher. 2003. "Community Development in Sustainable Livelihoods Approaches - an Introduction". *Community Development Journal* 38, no. 3: 185–98. https://doi.org/10.1093/cdj/38.3.185.

Brown, David. 2019. "Climate Justice and REDD+: A Multiscalar Examination of the Norwegian-Ethiopian Partnership". In *Routledge Handbook of Climate Justice*, edited by Tahseen Jafry, pp. 262–75. London: Routledge.

Bruun, Ole. 2020. "Environmental Protection in the Hands of the State: Authoritarian Environmentalism and Popular Perceptions in Vietnam". *The Journal of Environment & Development* 29, no. 2: 171–95. https://doi.org/10.1177/1070496520905625.

Bui, Thiem Hai. 2013. "The Development of Civil Society and Dynamics of Governance in Vietnam's One Party Rule". *Global Change, Peace & Security* 25, no. 1: 77–93. https://doi.org/10.1080/14781158.2013.758100.

Callicott, J. Baird, ed. 1989. *Nature in Asian Traditions of Thought: Essays in Environmental Philosophy*. Albany: State University of New York Press.

Cannon, Terry, John Twigg, and Jennifer Rowell. 2003. *Social Vulnerability, Sustainable Livelihood and Disasters*. Report to DFID – Conflict and Humanitarian Assistance (CHAD) and Sustainable Livelihoods Support Office. Brighton: Institute of Development Studies UK.

Cha, J. Mijin. 2019. "From the Dirty Past to the Clean Future: Addressing Historic Energy Injustice with a Just Transition to a Low-Carbon Future". In *Routledge Handbook of Climate Justice*, edited by Tahseen Jafry, pp. 211–21. London: Routledge.

Chambers, Robert. 1998. *Whose Reality Counts? Putting the First Last*. London: Intermediate Technology.

Chang, Yufen. 2016. "Traveling Civilization: The Sinographic Translation Network and Modern Lexicon Building in Colonial Vietnam, 1890s–1910s". Nalanda-Sriwijaya Centre Working Paper 21. Singapore: ISEAS – Yusof Ishak Institute.

Cho, Sumi, Kimberlé Williams Crenshaw, and Leslie McCall. 2013. "Toward a Field of Intersectionality Studies: Theory, Applications and Praxis". *Journal of Women in Culture and Society* 38, no. 4: 785–810.

Cole, Robert, and Micah L. Ingalls. 2020. "Rural Revolutions: Socialist, Market and Sustainable Development of the Countryside in Vietnam and Laos". In *The Socialist Market Economy in Asia*, edited by Jo I. Bekkevold, Arve Hansen, and Kristen Nordhaug, pp. 167–94. Basingstoke: Palgrave Macmillan.

Cuc, Le Trong. 1999. "Vietnam: Traditional Cultural Concepts of Human Relations with the Natural Environment". *Asian Geographer* 18, no. 1–2: 67–74. https://doi.org/10.1080/10225706.1999.9684048.

Dahm, Bernhard, Vincent J. H. Houben, Martin Großheim, Kirsten W. Endres, and Annette Spitzenpfeil. 1999. *Vietnamese Villages in Transition*. Passau, Germany: Department of Southeast Asian Studies, University of Passau.

De Koninck, Rodolphe. 1998. "La Logique De La Déforestation En Asie Du Sud-Est" [The logic of deforestation in Southeast Asia]. *caoum* 51, no. 204: 339–66. https://doi.org/10.3406/caoum.1998.3702.

Degenhardt, Philip. 2016. *From Sustainable Development to Socio-ecological Transformation: An Overview*. Berlin: Rosa-Luxemburg-Stiftung.

Dellheim, Judith, and Guenter Krause eds. 2008. *Fuer Eine Neue Alternative: Herausforderungen Einer Sozialoekologischen Transformation*. With the assistance of Guenter Krause. Manuskripte. Berlin: Karl Dietz.

Derks, Annuska, and Minh T. N. Nguyen. 2020. "Beyond the State? The Moral Turn of Development in South East Asia". *South East Asia Research* 28, no. 1: 1–12.

Derman, Brandon. 2019. "'Climate Change Is About Us': Fence-Line Communities, the NAACP and the Grounding of Climate Justice". In *Routledge Handbook of Climate Justice*, edited by Tahseen Jafry, pp. 407–19. London: Routledge.

Déry, Steve. 2000. "Agricultural Colonisation in Lam Dong Province, Vietnam". *Asia Pacific Viewpoint* 41, no. 1: 35–49. https://doi.org/10.1111/1467-8373.00105.

Déry, Steve, and Romain Vanhooren. 2011. "Protected Areas in Mainland Southeast Asia, 1973–2005: Continuing Trends". *Singapore Journal of Tropical Geography* 32, no. 2: 185–202. https://doi.org/10.1111/j.1467-9493.2011.00428.x.

Devos, Jean Claude, Jean Nicot, and Philippe Schillinger. 1990. *Inventaire dé Archives de l'Indochine: Sous serie 10H (1867–1956)* [Inventory of the archives of Indochina: Sub-series 10H (1867–1956)]. Châuteau de Vincennes: Service Historique de l'armee de terre.

Dewan, Camelia. 2020. "'Climate Change as a Spice': Brokering Environmental Knowledge in Bangladesh's Development Industry". *Ethnos* 87, no. 3: 538–59. https://doi.org/10.1080/00141844.2020.1788109

———. 2021. *Misreading the Bengal Delta: Climate Change, Development, and Livelihood in Coastal Bangladesh*. Seattle: University of Washington Press.

DiGregorio, Michael, and Oscar Salemink. 2007. "Living with the Dead: The Politics of Ritual and Remembrance in Contemporary Vietnam". *Journal of Southeast Asian Studies* 38, no. 3: 433–40.

DiGregorio, Monica, Maria Brockhaus, Tim Cronin, Efrian Muharrom, Sofi Mardiah, and Levania Santoso. 2015. "Deadlock or Transformational Change? Exploring Public Discourse on REDD+ Across Seven Countries". *Global Environmental Politics* 15, no. 4: 63–84. https://doi.org/10.1162/GLEP_a_00322.

Doherty, Brian, and Timothy Doyle. 2018. "Friends of the Earth International: Agonistic Politics, Modus Vivendi and Political Change". *Environmental Politics* 27, no. 6: 1057–78. https://doi.org/10.1080/09644016.2018.1462577.

Doyle, Timothy, and Adam Simpson. 2007. "Traversing More than Speed Bumps: Green Politics under Authoritarian Regimes in Burma and Iran". *Environmental Politics* 15, no. 5: 750–67.

Dung, Thuy. 2023. "Business Sector to Contribute 65-70% of Viet Nam's GDP by 2025". *Government News Socialist Republic of Vietnam*, 22 April 2023. https://en.baochinhphu.vn/business-sector-to-contribute-65-70-of-viet-nams-gdp-by-2025-111230422095757233.htm (accessed 3 March 2024).

Duong, Van Ni, Roger Safford, and Edward Maltby. 2001. "Environmental Change, Ecosystem Degradation and the Value of Wetland Rehabilitation in the Mekong Delta". In *Living with Environmental Change*, edited by W. Neil Adger, P. Mick Kelly, and Nguyen Huu Ninh, pp. 122–36. London: Routledge.

Easterly, William. 2014. *The Tyranny of Experts: Economists, Dictators, and the Forgotten Rights of the Poor*. New York: Basic Books.

Edelman, Marc, and Angelique Haugerud. 2005. *The Anthropology of Development and Globalization: From Classical Political Economy to Contemporary Neoliberalism*. Blackwell anthologies in social and cultural anthropology 6. Oxford: Blackwell Publishing Ltd.

Eggers, Maureen M., ed. 2009. *Mythen, Masken Und Subjekte: Kritische Weissseinsforschung in Deutschland* [*Myths, masks and subjects: Critical whiteness research in Germany*]. 2, überarbeitete Aufl. Münster: Unrast.

Escobar, Arturo. 1995. *Encountering Development: The Making and Unmaking of the Third World*. Princeton, NJ: Princeton University Press.

Falkner, Robert, ed. 2013. *The Handbook of Global Climate and Environment Policy*. Oxford: Wiley-Blackwell.

Fan, Mei-Fang, Chih-Ming Chiu, and Leslie Mabon. 2020. "Environmental Justice and the Politics of Pollution: The Case of the Formosa Ha Tinh Steel Pollution Incident in Vietnam". *Environment and Planning E: Nature and Space* 5, no. 1. https://doi.org/10.1177/2514848620973164.

Fatheuer, Thomas, Lili Fuhr, and Barbara Unmüssig. 2015. *Kritik Der Grünen Ökonomie* [Critique of the green economy]. München: Oekom Verlag.

Ferguson, James. 1994. *The Anti-politics Machine: Development, Depoliticization, and Bureaucratic Power in Lesotho*. Minneapolis: University of Minnesota Press.

———. 2005. "Anthropology and Its Evil Twin: 'Development' in the Constitution of a Discipline". In *The Anthropology of Development and Globalization*, edited by Marc Edelmann and Angelique Haugerud, pp. 140–54. Malden: Blackwell Publishing.

Folke Ax, Christina, ed. 2011. *Cultivating the Colonies: Colonial States and Their Environmental Legacies*. Athens: Ohio University Press.

Forsyth, Tim, and Andrew Walker. 2008. *Forest Guardians, Forest Destroyers: The Politics of Environmental Knowledge in Northern Thailand*. Seattle, WA: University of Washington Press.

Fortier, François. 2010. "Taking a Climate Chance: A Procedural Critique of Vietnam's Climate Change Strategy". *Asia Pacific Viewpoint* 51, no. 3: 229–47.

Foster, John B. 1999. *Marx's Ecology: Materialism and Nature*. New York: Monthly Review Press.

Freud, Benjamin. 2014. "Organizing Autarky: Governor General Decoux's Development of a Substitution Economy in Indochina as a Means of Promoting Colonial Legitimacy". *Journal of Social Issues in Southeast Asia* 29, no. 1: 96–131.

Fuenfgeld, Anna. 2019. "Just Energy? Structures of Energy (In)Justice and the Indonesian Coal Sector". In *Routledge Handbook of Climate Justice*, edited by Tahseen Jafry, pp. 222–36. London: Routledge.

Fuhrmann, Eva. 2017. *Perceptions of Change in Vietnam*. Berlin: regiospectra.

Fukuyama, Francis. 1992. *The End of History and the Last Man*. New York: Free Press.

Gainsborough, Martin. 2010. "Present but Not Powerful: Neoliberalism, the State, and Development in Vietnam". *Globalizations* 7, no. 4: 475–88. https://doi.org/10.1080/14747731003798435.

Glaser, Barney, and Anselm Strauss. 2017. *Discovery of Grounded Theory: Strategies for Qualitative Research*. London: Routledge.

Global Energy Monitor. 2022. *Global Coal Plant Tracker*. https://globalenergymonitor.org/projects/global-coal-plant-tracker/tracker (accessed 20 August 2022).

Gonzalez, Carmen. 2000. "Beyond Eco-imperialism: An Environmental Justice Critique of Free Trade". *Denver University Law Review* 78: 979–1016.

Goscha, Christopher E. 2016. *The Penguin History of Modern Vietnam*. London: Penguin.

Grant, Wyn. 2000. *Pressure Groups and British Politics*. Basingstoke: Palgrave Macmillan.

Grossheim, Martin. 2004. "Village Government in Pre-Colonial and Colonial Vietnam". In *Beyond Hanoi: Local Government in Vietnam*, edited by Benedict Kerkvliet, pp. 54–89. Copenhagen: NIAS Press.

Grunwald, Armin. 2011. "Sustainable Development. An Issue of the Office of Technology Assessment at the German Bundestag". In *Sustainable Development - the Cultural Perspective: Concepts - Aspects - Examples*, edited by Gerhard Banse, Gordon L. Nelson, and O. Parodi, pp. 125–36. Gesellschaft, Technik, Umwelt. Berlin: Edition Sigma.

Hajer, Maarten A. 1995. *The Politics of Environmental Discourse: Ecological Modernization and the Policy Process*. Oxford: Clarendon Press.

Hakkarainen, Minna. 2012. "The Role of Interpretations of 'Participation' in Development Practice". In *Perspectives on Difference: Makings and Workings of Power*, edited by Anni Kajanus, vol. 30, pp. 1–28. Helsinki: Renvall Institute.

Hannah, Joseph. 2007. "Local Non-Governmental Organizations in Vietnam: Development, Civil Society and State-Society Relations". PhD thesis, University of Washington.

Hansen, Mette Halskov, Hongtao Li, and Rune Svarverud. 2018. "Ecological Civilization: Interpreting the Chinese Past, Projecting the Global Future". *Global Environmental Change* 53: 195–203. https://doi.org/10.1016/j.gloenvcha.2018.09.014.

Harms, Erik. 2012. "Beauty as Control in the New Saigon: Eviction, New Urban Zones, and Atomized Dissent in a Southeast Asian City". *American Ethnologist* 39, no. 4: 735–50. https://doi.org/10.1111/j.1548-1425.2012.01392.x.

———. 2016. *Luxury and Rubble: Civility and Dispossession in the New Saigon*. Oakland, California: University of California Press.

Healy, Noel, and John Barry. 2017. "Politicizing Energy Justice and Energy System Transitions: Fossil Fuel Divestment and a 'Just Transition'". *Energy Policy* 108: 451–59. https://doi.org/10.1016/j.enpol.2017.06.014.

Hiep, Le Hong. 2020. "Previewing Vietnam's Leadership Transition in 2021". *ISEAS Perspective*, no. 2020/41, 8 May 2020.

Holzhacker, Ronald, and Dafri Agussalim. 2019. "Introduction: Sustainable Development Goals in Southeast Asia and ASEAN". In *Sustainable Development Goals in Southeast Asia and ASEAN*, edited by Ronald Holzhacker and Dafri Agussalim, pp. 3–38. Leiden: Brill.

———, eds. 2019. *Sustainable Development Goals in Southeast Asia and ASEAN*. Leiden: Brill.

Hong, Phan Nguyen, and Hoang Thi San. 1993. *Mangroves of Vietnam*. Bangkok: IUCN.

Institute of Strategy and Policy on Natural Resources and Environment (ISPONRE) Vietnam. 2009. *Vietnam Assessment Report on Climate Change*. Hanoi: Kim Do Publishing House.

Jacob, Klaus, Philipp Kauppert, and Rainer Quitzow. 2013. *Green Growth Strategies in Asia: Drivers and Political Entry Points*. Berlin: Friedrich-Ebert-Stiftung.

Jacobs, Michael. 2013. "Green Growth". In *The Handbook of Global Climate and Environment Policy*, edited by Robert Falkner, pp. 197–224. Oxford: Wiley-Blackwell.

Jafry, Tahseen. 2019. "Introduction: Justice in the Era of Climate Change". In *Routledge Handbook of Climate Justice*, edited by Tahseen Jafry, pp. 1–9. London: Routledge.

Jafry, Tahseen, Karin Helwig, and Michael Mikulewicz, eds. 2019. *Routledge Handbook of Climate Justice*. London: Routledge.

James, Simon P., and David E. Cooper. 2007. "Buddhism and the Environment". *Contemporary Buddhism* 8, no. 2: 93–96. https://doi.org/10.1080/14639940701636075.

Jamieson, Neil L. 1995. *Understanding Vietnam*. Berkeley: University of California Press.

Jiang Chunyun. 2013. "Chinese Leader Calls for Ecological Civilization: Climate and Capitalism". *Climate & Capitalism*, 31 March 2013. https://climateandcapitalism.com/2013/03/31/chinese-leader-calls-for-ecological-civilization/ (accessed 6 September 2021).

Keck, Margaret E., and Kathryn Sikkink. 1998. *Activists Beyond Borders: Advocacy Networks in International Politics*. Ithaca, NY: Cornell University Press.

Kerkvliet, Benedict J. Tria. 2001. "Analysing the State in Vietnam". *Sojourn* 16, no. 2: 179–86.

———. 2005. *The Power of Everyday Politics: How Vietnamese Peasants Transformed National Policy*. Ithaca, NY: Cornell University Press.

———. 2019. *Speaking Out in Vietnam: Public Political Criticism in a Communist Party-Ruled Nation*. Ithaca, NY: Cornell University Press.

———, ed. 2004. *Beyond Hanoi: Local Government in Vietnam*. Copenhagen: NIAS Press.

Kerkvliet, Benedict J. Tria, Russell Hiang-Khng Heng, and David W. H. Koh, eds. 2003. *Getting Organized in Vietnam*. Singapore: ISEAS – Yusof Ishak Institute.

Kerkvliet, Benedict J. Tria, and David G. Marr. 2004. *Beyond Hanoi: Local Government in Vietnam*. Singapore: ISEAS – Yusof Ishak Institute.

Kerkvliet, Benedict J. Tria, and Doug Porter, eds. 1995. *Vietnam's Rural Transformation (Transitions: Asia and Asian America)*. Boulder, CO: Westview Press.

Kilomba, Grada. 2009. "No Mask". In *Mythen, Masken und Subjekte: Kritische Weißseinsforschung in Deutschland*, edited by Maureen Maisha Eggers, Garda Kilomba, Peggy Pirsche, and Susan Arndt. Münster: Unrast.

Latour, Bruno. 1993. *We Have Never Been Modern*. Cambridge, MA: Harvard University Press.

———. 2007. *Reassembling the Social: An Introduction to Actor-Network-Theory*. Oxford: Oxford University Press.

———. 2017. *Kampf um Gaia: acht Vorträge über das neue Klimaregime*. Berlin: Suhrkamp Verlag.

Lauser, Andrea. 2008. "Zwischen Heldenverehrung Und Geisterkult: Politik Und Religion Im Gegenwärtigen Spiritkommunismus Vietnam". *Zeitschrift fuer Ethnologie* 133: 121–44.

Leshkowich, Ann Marie. 2008. "Wandering Ghosts of Late Socialism: Conflict, Metaphor, and Memory in a Southern Vietnamese Marketplace". *The Journal of Asian Studies* 67, no. 1: 5–41.

Levine, Simone. 2014. "How to Study Livelihoods: Bringing a Sustainable Livelihoods Framework to Life". Secure Livelihoods Research Consortium (SLRC) Working Paper 22. London: Overseas Development Institute.

Li, Tania. 2005. "Beyond 'The State' and Failed Schemes". *American Anthropologist* 107, no. 3: 383–94. https://doi.org/10.1525/aa.2005.107.3.383.

———. 2007. *The Will to Improve: Governmentality, Development, and the Practice of Politics*. Durham, NC: Duke University Press.

Lievens, Matthias. n.d. *Towards an Eco-Marxism*. Belgium: KU Leuven.

Lindegaard, Lily S. 2020. "Global Climate Change Knowledge and the Production of Climate Subjects in Vietnam". *Forum for Development Studies* 47, no. 1: 157–80. https://doi.org/10.1080/08039410.2019.1685590.

London, Jonathan D., ed. 2014. *Politics in Contemporary Vietnam*. London: Palgrave Macmillan UK.

Lundberg, Mats. 2004. "Kinh Settlers in Viet Nam's Northern Highlands: Natural Resources Management in a Cultural Context". PhD thesis, University of Linkoeping.

Luttrell, Cecilia. 2001. "Institutional Change and Natural Resource Use in Coastal Vietnam". *GeoJournal* 55: 529–40.

MacLeod, Roy, ed. 2000. *Nature and Empire: Science and the Colonial Enterprise*. Chicago: University of Chicago Press.

Magnusson, Eva, and Jeanne Marecek. 2015. *Doing Interview-Based Qualitative Research: A Learner's Guide*. Cambridge: Cambridge University Press.

Mahanty, Sango, and Trung Dinh Dang. 2015. "Between 'State' and 'Society': Commune Authorities and the Environment in Vietnam's Craft Villages". *Asia Pacific Viewpoint* 56, no. 2: 267–81. https://doi.org/10.1111/apv.12077.

Marr, David. 2004. "A Brief History of Local Government in Vietnam". In *Beyond Hanoi: Local Government in Vietnam*, edited by Benedict Kerkvliet, pp. 28–53. Copenhagen: NIAS Press.

Mayrhofer, Jan P., and Joyeeta Gupta. 2016. "The Science and Politics of Co-benefits in Climate Policy". *Environmental Science & Policy* 57: 22–30.

McCauley, Darren, and Raphael Heffron. 2018. "Just Transition: Integrating Climate, Energy and Environmental Justice". *Energy Policy* 119: 1–7. https://doi.org/10.1016/j.enpol.2018.04.014.

McClellan III, James E. 2000. "The Colonial Machine: French Science and Colonialization in the Ancient Regime". In *Nature and Empire: Science and the Colonial Enterprise*, edited by Roy MacLeod, pp. 31–50. Chicago: University of Chicago Press.

McHale, Shawn Frederick. 2004. *Print and Power*. Honolulu: University of Hawaii Press.

McElwee, Pamela D. 2012. "Payments for Environmental Services as Neoliberal Market-Based Forest Conservation in Vietnam: Panacea or Problem?" *Geoforum* 43: 412–26.

———. 2016. *Forests Are Gold: Trees, People, and Environmental Rule in Vietnam*. Seattle: University of Washington Press.

McGregor, Callum, Eurig Scandrett, Elizabeth Christie, and James Crowther. 2018. "Climate Justice Education: From Social Movement Learning to Schooling". In *Routledge Handbook of Climate Justice*, edited by Tahseen Jafry. London: Routledge.

McManus, Phil, and Graham Haughton. 2006. "Planning with Ecological Footprints: A Sympathetic Critique of Theory and Practice". *Environment and Urbanization* 18, no. 1: 113–27.

Micheaud, Jean. 2015. "Livelihoods in the Vietnamese Northern Borderlands Recorded in French Colonial Military Ethnographies 1897–1904". *The Asia Pacific Journal of Anthropology* 16, no. 4: 343–67. https://doi.org/10.1080/14442213.2015.1044559.

Mielke, Katja, and Anna-Katharina Hornidge, eds. 2017. *Area Studies at the Crossroads: Knowledge Production after the Mobility Turn*. New York, NY: Springer Science.

Mol, Arthur P.J. 2003. *Globalization and Environmental Reform: The Ecological Modernization of the Global Economy*. Cambridge: MIT Press.

Morris, Christopher. 2011. "Wetland Colonies: Louisiana, Guangzhou, Pondicherry and Senegal". In *Cultivating the Colonies: Colonial States and Their Environmental Legacies*, edited by Christina Folke Ax, pp. 135–64. Athens: Ohio University Press.

Mosse, David. 2005. *Cultivating Development: An Ethnography of Aid Policy and Practice*. London: Pluto Press.

Murphy, Susan P. 2019. "Global Political Processes and the Paris Agreement: A Case of Advancement or Retreat of Climate Justice?" In *Routledge Handbook of Climate Justice*, edited by Tahseen Jafry, pp. 71–82. London: Routledge.

Natarajan, Usha. 2021. "Environmental Justice in the Global South". In *The Cambridge Handbook of Environmental Justice and Sustainable Development*, edited by Sumudu A. Atapattu, Carmen G. Gonzalez, and Sara L. Seck, pp. 39–57. Cambridge: Cambridge University Press.

Newell, Peter, and Dustin Mulvaney. 2013. "The Political Economy of the 'Just Transition'". *The Geographical Journal* 179, no. 2: 132–40. https://doi.org/10.1111/geoj.12008.

Nguyen, Chi Quoc. 2012. *Greening of Doi Moi: An Outlook on the Potential of Green Jobs in Vietnam*. Hanoi: Friedrich-Ebert-Stiftung.

Nguyen, Minh T. N. 2014. *Vietnam's Socialist Servants: Domesticity, Class, Gender, and Identity*. London: Routledge.

―――. 2018. "Vietnam's 'Socialization' Policy and the Moral Subject in a Privatizing Economy", *Economy and Society* 47, no. 4: 627–47.

Nguyen, Nguyet Cam, and Dana Sachs. 2003. *Two Cakes Fit for a King: Folktales from Vietnam*. Honolulu: University of Hawaii Press.

Nguyễn, Phú Trọng. 2021. "Some Theoretical and Practical Issues on Socialism and Path towards Socialism in Vietnam". *Vietnam Law Magazine*, 17 May 2021. https://vietnamlawmagazine.vn/some-theoretical-and-practical-issues-on-socialism-and-path-towards-socialism-in-vietnam-37726.html (accessed 28 July 2024).

Nguyen, Quang Dung. 2020. *From Cyberspace to the Streets: Emerging Environmental Paradigm of Justice and Citizenship in Vietnam*. Leiden: International Institute for Asian Studies.

Noblit, George W., Susana Y. Flores, and Enrique G. Murillo, eds. 2004. *Postcritical Ethnography: Reinscribing Critique (Understanding Education and Policy)*. Cresskill, NJ: Hampton Press.

Ogawa, Akihiro, ed. 2018. *Routledge Handbook of Civil Society in Asia*. London: Routledge.

Ong, Aihwa, and Li Zhang. 2008. *Privatizing China: Socialism from Afar*. Ithaca: Cornell University Press.

Ophuls, William. 1973. *"Leviathan or Oblivion" towards a Steady State Economy*. New York: W. H. Freeman.

O'Rourke, Dara. 2004. *Community-Driven Regulation: Balancing Development and the Environment in Vietnam*. London: MIT Press.

Ortmann, Stephan. 2017. *Environmental Governance in Vietnam*. Cham: Springer International Publishing.

―――. 2020. "Evolving Environmental Governance Structures in a Market Socialist State: The Case of Vietnam". In *The Socialist Market Economy in Asia: Development in China, Vietnam and Laos*, edited by Arve Hansen, Jo Inge Bekkevold, and Kristen Nordhaug, pp. 195–217. Basingstoke: Palgrave Macmillan.

―――. 2021. "Environmental Movements in Vietnam under One-Party Rule". In *Environmental Movements and Politics of the Asian Anthropocene*, edited by Paul Jobin, Ming-sho Ho, and Hsin-Huang Michael Hsiao, pp. 261–93. Singapore: ISEAS – Yusof Ishak Institute.

Page, Edward. 2013. "Climate Change Justice". In *The Handbook of Global Climate and Environment Policy*, edited by Robert Falkner, pp. 231–48. Oxford: Wiley-Blackwell.

Painter, Martin. 2006. "Sequencing Civil Service Pay Reforms in Vietnam: Transition or Leapfrog?" *Governance* 19, no. 2: 325–47. https://doi.org/10.1111/j.1468-0491.2006.00317.x.

Pearce, David, Anil Markandya, and Edward Barbier. 2000. *Sustainable Development: Economics and Environment in the Third World*. London: Routledge.

Perkins, Patricia E. 2019. "Climate Justice, Gender and Intersectionality". In *Routledge Handbook of Climate Justice*, edited by Tahseen Jafry, pp. 349–58. London: Routledge.

Persoon, Gerard A. 1997. *Defining Wildness and Wilderness: Minangkabau Images and Actions on Siberut (West Sumatra)*. Canterbury, Brussels: APFT.

Petras, James. 1999. "NGOs: In the Service of Imperialism". *Journal of Contemporary Asia* 29, no. 4: 429–40. https://doi.org/10.1080/00472339980000221.

Pham, Quang Minh. 2004. "Caught in the Middle: Local Cadres in Hai Duong Province". In *Beyond Hanoi: Local Government in Vietnam*, edited by Benedict Kerkvliet, pp. 90–109. Copenhagen: NIAS Press.

―――. 2021. Interview by Julia Behrens. Hanoi. June 2021. https://suedostasien.net/vietnam-interview-prinzip-der-gleichheit-und-gegenseitigkeit/(accessed 16 September 2021).

Pham, Thi Thuong Vi, and A. Terry Rambo. 2003. "Environmental Consciousness in Vietnam". *Japanese Journal of Southeast Asian Studies* 41, no. 1: 76–100.

Pham, Thu Thuy, Bruce M. Campbell, Stephen Garnett, Heather Aslin, and Minh Ha Hoang. 2010. "Importance and Impacts of Intermediary Boundary Organizations in Facilitating Payment for Environmental Services in Vietnam". *Environmental Conservation* 37, no. 1: 64–72. https://doi.org/10.1017/S037689291000024X.

Pham Thi, Anh-Susann. 2020. "Activism in Vietnam: Political Practice and Cognitive Resistance". PhD thesis, University of Manchester.

Pham Thi, Anh-Susann, and Julia Behrens. 2019. "Autoritaerer Staat vs. solidarische Zivilgesellschaft: Reaktionen auf Umweltproteste in Vietnam" [Authoritarian State vs. Civil Society of Solidarity: Reactions to Environmental Protests in Vietnam]. *Suedostasien*. https://suedostasien.net/author/julia-behrens-anh-susann-pham-thi/ (accessed 3 March 2020).

Polanyi, Karl. 1978. *The Great Transformation: The Political and Economic Origins of Our Time*. Boston, MA: Beacon Press.

Porter, Gareth. 1993. *Vietnam: The Politics of Bureaucratic Socialism*. Ithaca: Cornell University Press.

Princen, Thomas, and Matthias Finger. 1994. *Environmental NGOs in World Politics*. London: Routledge.

Quitzow, Rainer, Holger Baer, and Klaus Jacob. 2013. "Environmental Governance in India, China, Vietnam and Indonesia: A Tale of Two Paces". FFU-Report 01-2013. FU Berlin: Forschungszentrum für Umweltpolitik.

Rademacher, Anne, and K. Sivaramakrishnan, eds. 2017. *Places of Nature in Ecologies of Urbanism*. Hong Kong: Hong Kong University Press.

Rambo, A. Terry. 2006. "Human Ecology Research on Tropical Agrosystems in Southeast Asia". *Singapore Journal of Tropical Geography* 3, no. 1: 86–99. https://doi.org/10.1111/j.1467-9493.1982.tb00232.x.

Report of the World Commission on Environment and Development. 1987. *Our Common Future*. Report presented to the United Nations.

Rosemberg, Anabella. 2010. "Building a Just Transition: The Linkages between Climate Change and Employment". *International Journal of Labour Research* 2, no. 2: 126–61.

Roubequain, Charles. 1939. *Centre D'etudes De Politique Etrangere: Travaux Des Groupes D'etudes* [Centre of Foreign Policy Studies: Works of the Study Group]. Paris: P. Hartmann.

Rumsby, Seb. 2020. "Alternative Routes to Development? The Everyday Political Economy of Christianisation among a Marginalised Ethnic Minority in Vietnam's Highlands". PhD thesis, University of Warwick.

Sachs, Wolfgang, ed. 1992. *The Development Dictionary: A Guide to Knowledge as Power*. London: Zed Books.

Saed. 2019. "James Richard O'Connor's Ecological Marxism". *Capitalism Nature Socialism* 30, no. 4: 1–12. https://doi.org/10.1080/10455752.2018.1495307.

Said, Edward W. 1978. *Orientalism*. New York: Pantheon Books.

Saito, Kohei. 2023. *Marx in the Anthropocene: Towards the Idea of Degrowth Communism*. Cambridge: Cambridge University Press.

Salemink, Oscar. 2011. "A View from the Mountains: A Critical History of Lowlander–Highlander Relations in Vietnam". In *Upland Transformations in Vietnam*, edited by Thomas Sikor, Nghiem Phuong Tuyen, Jennifer Sowerwine, and Jeff Romm, pp. 27–50. Singapore: NUS Press.

Sassen, Saskia. 2008. *Territory, Authority, Rights: From Medieval to Global Assemblages*. Princeton: Princeton University Press.

———. 2014. *Expulsions: Brutality and Complexity in the Global Economy*. Cambridge: Harvard University Press.

Schirmbeck, Sonja. 2017. *Vietnam's Environmental Policies at a Crossroads: Salinated Rice Fields, Hunted-Out National Parks, and Eroding Beaches - and What We Can Do about It*. Hanoi: Friedrich-Ebert-Stiftung.

Schlosberg, David. 2004. "Reconceiving Environmental Justice: Global Movements and Political Theories". *Environmental Politics* 13, no. 3: 517–40.

———. 2007. *Defining Environmental Justice: Theories, Movements, and Nature*. Oxford: University of Oxford Press.

———. 2013. "Theorising Environmental Justice: The Expanding Sphere of a Discourse". *Environmental Politics* 22, no. 1: 37–55. https://doi.org/10.1080/09644016.2013.755387.

Schreurs, Miranda A., and Elizabeth E. Economy. 1997. *The Internalization of Environmental Protection*. Cambridge: Cambridge University Press.

Schwenkel, Christina. 2017. "Eco-Socialism and Green City Making in Postwar Vietnam". In *Places of Nature in Ecologies of Urbanism*, edited by Anne Rademacher and K. Sivaramakrishnan, pp. 45–66. Hong Kong: Hong Kong University Press.

———. 2020. *Building Socialism: The Afterlife of East German Architecture in Urban Vietnam*. Durham: Duke University Press.

Schwenkel, Christina, and Marie Leshkowich. 2012. "How Is Neoliberalism Good to Think Vietnam? How Is Vietnam Good to Think Neoliberalism?" *Positions* 20, no. 2: 381–401. https://doi.org/10.1215/10679847-1538461.

Scoones, Ian. 2009. "Livelihoods Perspectives and Rural Development". *The Journal of Peasant Studies* 36, no. 1: 171–96. https://doi.org/10.1080/03066150902820503.

Scott, James C. 1985. *Weapons of the Weak: Everyday Forms of Peasant Resistance*. New Haven, CT: Yale University Press.

———. 1998. *Seeing Like a State: How Certain Schemes to Improve the Human Condition Have Failed*. New Haven, CT: Yale University Press.

———. 2009. *The Art of Not Being Governed: An Anarchist History of Upland Southeast Asia*. New Haven, CT: Yale University Press.

Serrat, Olivier. 2017. "The Sustainable Livelihoods Approach". In *Knowledge Solutions*, edited by Olivier Serrat, pp. 21–26. Singapore: Springer Singapore.

_____, ed. 2017. *Knowledge Solutions*. Singapore: Springer Singapore.

Sheyvens, Regina, and Helen Leslie. 2000. "Gender, Ethics and Empowerment: Dilemmas of Development Fieldwork". *Women's Studies International Forum* 23, no. 1: 119–30.

Sikkink, Kathryn. 1993. "Human Rights, Principled Issue-Networks, and Sovereignty in Latin America". *Int Org* 47, no. 3: 411–41.

Sikor, Thomas. 2013. "REDD+: Justice Effects of Technical Design". In *The Justices and Injustices of Ecosystem Services*, edited by Thomas Sikor, pp. 46–68. London: Routledge.

Sikor, Thomas, Nghiem Phuong Tuyen, Jennifer Sowerwine, and Jeff Room, eds. 2011. *Upland Transformations in Vietnam*. Singapore: NUS Press.

Sivaramakrishnan, K. 1995. "Situating the Subaltern: History and Anthropology in the Subaltern Studies Project". *Journal of Historical Sociology* 8, no. 4: 395–429.

Sivaramakrishnan, K., and Arun Agrawal, eds. 2003. *Regional Modernities: The Cultural Politics of Development in India*. Oxford: Oxford University Press.

Skirbekk, Gunnar. 1994. "Marxism and Ecology". *Capitalism, Nature, Socialism* 5, no. 4: 95–104.

Smith, Alison. 2013. *The Climate Bonus: Co-benefits of Climate Policy*. London: Routledge.

Spiegel, Anna. 2010. "Entering the World of NGOs: The Researcher's Trajectory". In *Contested Public Spheres VS Verlag für Sozialwissenschaften*, pp. 33–57. https://link.springer.com/chapter/10.1007/978-3-531-92371-0_2.

Spoor, Max, Nico Heerink, and Futian Qu, eds. 2007. *Dragons with Clay Feet? Transition, Sustainable Land Use, and Rural Environment in China and Vietnam*. Lanham, MD: Lexington Books.

Swilling, Mark. 2019. *The Age of Sustainability: Just Transitions in a Complex World*. London: Routledge.

Taenzler, Dirk, Konstadinos Maras, and Angelos Giannakopoulos. 2012. *The Social Construction of Corruption in Europe*. Burlington, VT: Ashgate.

Tamoudi, Nejma, and Michael Reder. 2019. "A Narrative Account of Temporality in Climate Justice". In *Routledge Handbook of Climate Justice*, edited by Tahseen Jafry, pp. 57–68. London: Routledge.

Tänzler, Dirk, Konstadinos Maras, and Angelos Giannakopoulos. 2012. *The Social Construction of Corruption in Europe*. Burlington, VT: Ashgate.

Thayer, Carlyle A. 2014. "The Apparatus of Authoritarian Rule in Vietnam". In *Politics in Contemporary Vietnam*, edited by Jonathan D. London, pp. 135–61. London: Palgrave Macmillan UK.

Thomas, Frederic. 2009. "Protection Des Forets Et Environnementalisme Colonial: Indochine, 1860–1945" [Forest Protection and Colonial Environmentalism: Indochina, 1860–1945]. *Revue d'histoire moderne & contemporaine* 56–4, no. 4: 104–36.

To, Kien. 2011. "Cultural Sustainability and Sustainable Community Initiative in Developing Countries: Evidence from Vietnam and Indonesia". In *Sustainable Development - the Cultural Perspective: Concepts - Aspects - Examples*, edited by Gerhard Banse, Gordon L. Nelson, and O. Parodi, pp. 349–76. Gesellschaft, Technik, Umwelt. Berlin: Edition Sigma.

Tran, Thi Minh Thi. 2021. "Complex Transformation of Divorce in Vietnam under the Forces of Modernization and Individualism". *International Journal of Asian Studies* 18, no. 2: 225–45. https://doi.org/10.1017/S1479591421000115.

Transparency International. 2021. "Global Corruption Barometer EU: People Worried about Unchecked Abuses of Power". 15 June 2021. https://www.transparency.org/en/news/gcb-eu-2021-survey-people-worry-corruption- unchecked-impunity-business-politics.

Truong, Huyen Chi. 2004. "Winter Crop and Spring Festival: The Contestations of Local Government in a Red River Delta Commune". In *Beyond Hanoi: Local Government in Vietnam*, edited by Benedict Kerkvliet, pp. 110–36. Copenhagen: NIAS Press.

Tsing, Anna L. 2000. "The Global Situation". *Cultural Anthropology* 15, no. 3: 327–60. https://doi.org/10.1525/can.2000.15.3.327.

_____. 2005. *Friction: An Ethnography of Global Connection*. Princeton, NJ: Princeton University Press.

Tucker, Mary Evelyn, ed. 1998. *Confucianism and Ecology*. Cambridge: Harvard University Center for the Study of World Religions.

Tulbure, Ildiko. 2011. "Considerations regarding Cultural Differences When Operationalizing Sustainability on a Regional Level". In *Sustainable Development - the Cultural Perspective: Concepts - Aspects - Examples*, edited by Gerhard Banse, Gordon L. Nelson, and O. Parodi, pp. 125–36. Gesellschaft, Technik, Umwelt. Berlin: Edition Sigma.

UNDP (United Nations Development Programme) and MPI (Ministry of Planning and Investment). 1999. *A Study on Aid to the Environmental Sector in Vietnam*. Hanoi: UNDP and MPI.

UNFCCC (United Nations Framework Convention on Climate Change). n.d. "REDD+". *REDD+ Web Platform*. https://redd.unfccc.int/ (accessed 20 November 2021).

United Nations. n.d. "The 17 Goals". United Nations Department of Economic and Social Affairs Web Platform. https://sdgs.un.org/goals (accessed 4 September 2021).

United Nations General Assembly. 1992. *Report of the United Nations Conference on Environment and Development*. United Nations: Rio de Janeiro.

Ürge-Vorsatz, Diana, Sergio T. Herrero, Navroz K. Dubash, and Franck Lecocq. 2014. "Measuring the Co-benefits of Climate Change Mitigation". *Annual Review of Environment and Resources* 39, no. 1: 549–82. https://doi.org/10.1146/annurev-environ-031312-125456.

Vasavakul, Thaveeporn. 2003. "From Fence-Breaking to Networking: Interests, Popular Organizations, and Policy Influences in Post-Socialist Vietnam". In *Getting Organized in Vietnam*, edited by Benedict Kerkvliet, Russell Heng, and David Koh, pp. 25–61. Singapore: ISEAS – Yusof Ishak Institute.

⸺. 2019. *Vietnam*. Cambridge: Cambridge University Press.

Vo, Nguyen Giap. 2004. *Memoirs of War: The Road to Dien Bien Phu*. Hanoi: The Gioi.

Vu, Tuong. 2017. *Vietnam's Communist Revolution: The Power and Limits of Ideology*. Cambridge: Cambridge University Press.

VUSTA (Vietnam Union of Science and Technology Associations). 2018. "VUSTA Charter". VUSTA website. http://vusta.vn/chitieten/charter/charter-2018-11-19-10-9-940 (accessed 4 September 2021).

Watts, Michael J. 1993. "Development 1: Power, Knowledge, Discursive Practice". *Progress in Human Geography* 17, no. 2: 257–72.

Weger, Jacob. 2019. "The Vietnamization of Delta Management: The Mekong Delta Plan and Politics of Translation in Vietnam". *Environmental Science & Policy* 100: 183–88. https://doi.org/10.1016/j.envsci.2019.07.011.

Weller, Robert P. 2006. *Discovering Nature*. Cambridge: Cambridge University Press.

Wells-Dang, Andrew. 2011. "Informal Pathbreakers: Civil Society Networks in China and Vietnam". PhD thesis, University of Birmingham.

⸺. 2014. "The Political Influence of Civil Society in Vietnam". In *Politics in Contemporary Vietnam*, edited by Jonathan London, pp. 162–83. London: Palgrave Macmillan.

Wilcox, Pill, Jonathan Rigg, and Minh T.N. Nguyen. 2021. "Rural Life in Late Socialism: Politics of Development and Imaginaries of the Future". *European Journal of East Asian Studies* 20: 7–25.

Wischermann, Joerg. 2003. "Vietnam in the Era of *Doi Moi*: Issue-Oriented Organizations and Their Relationship to the Government". *Asian Survey* 43, no. 6: 867–889.

⸺. 2010. "Civil Society Action and Governance in Vietnam: Selected Findings from an Empirical Survey". *Journal of Current Southeast Asian Affairs* 29, no. 2: 3–40.

⸺. 2011. "Governance and Civil Society in Action in Vietnam: Changing the Rules from Within - Potentials and Limits". *Asian Politics and Policy* 3, no. 3: 383–411.

⸺. 2017. "Authoritarian Rule". In *Routledge Handbook of Civil Society in Asia*, edited by Akihiro Ogawa, pp. 344–61. London: Routledge.

⸺. 2018. "In Vietnam gibt es keine (wirkliche) Zivilgesellschaft!' Über einen politischen und wissenschaftlichen Mythos". In *Vietnam: Mythen und Wirklichkeiten*, edited by Jörg Wischermann and Gerhard Will, pp. 172–211. Bonn: Bundeszentrale für politische Bildung.

Wischermann, Joerg, and Dang Thi Viet Phuong. 2017. "Vietnam". In *Routledge Handbook of Civil Society in Asia*, edited by Akihiro Ogawa, pp. 129–42. London: Routledge.

Wischermann, Joerg, and Nguyen Quang Vinh. 2003. "The Relationship between Civic and Governmental Organizations in Vietnam: Selected Findings". In *Getting Organized in Vietnam*, edited by Ben J. T. Kerkvliet, Russell Hiang-Khng Heng, and David W. H. Koh, pp. 185–233. Singapore: ISEAS – Yusof Ishak Institute.

Wischermann, Joerg, and Gerhard Will, eds. 2018. *Vietnam: Mythen und Wirklichkeiten*. Schriftenreihe / Bundeszentrale für Politische Bildung Band 10297. Bonn: bpb Bundeszentrale für politische Bildung.

Wolf, Frieder Otto. 2008. "Umbau, Uebergang, Transformation: Arbeitsthesen Fuer Einen Begrifflichen Rahmen". In *Fuer Eine Neue Alternative: Herausforderungen Einer Sozialoekologischen Transformation*, edited by Judith Dellheim, pp. 25–46. Manuskripte. Berlin: Karl Dietz.

Wong, Wendy H. 2012. *Internal Affairs: How the Structure of NGOs Transforms Human Rights*. Ithaca, NY: Cornell University Press.

Workman, Annabelle, Grant Blashki, Kathryn J. Bowen, David J. Karoly, and John Wiseman. 2018. "The Political Economy of Health Co-benefits: Embedding Health in the Climate Change Agenda". *International Journal of Environmental Research and Public Health* 15, no. 4. https://doi.org/10.3390/ijerph15040674.

WWF Indochina Programme. 2002. *Johannesburg Project of Vietnam: Sustainable Development in Vietnam–a Report by Vietnamese NGOs*. Hanoi: WWF.

Yasuda, Yumiko. 2015. *Rules, Norms and NGO Advocacy Strategies: Hydropower Development on the Mekong River*. London: Routledge.

Zink, Eren. 2013. *Hot Science, High Water: Assembling Nature, Society and Environmental Policy in Contemporary Vietnam*. Copenhagen: NIAS Press.

Primary Sources

Government Sources

Central Committee of the Communist Party of Vietnam. 2013. Decision on the proactive response to climate change, strengthening resource management, and environmental protection 24NQ-TW.

Chair of the Cabinet of the Socialist Republic of Vietnam. 1991. Quyết định: về việc triển khai thực hiện kế hoạch quốc gia về môi trường và phát triển bền vững. 187-CT.

Government of the Socialist Republic of Vietnam. 2017. Resolution on Sustainable and Climate-Resilient Development of the Mekong Delta 120/NQ-CP.

———. 2019. Decree 93/2019/ND-CP. https://thuvienphapluat.vn/van-ban/EN/Tai-chinh-nha-nuoc/Decree-93-2019-ND-CP-prescribing-organization-and-operation-of-social-and-charity-funds/434363/tieng-anh.aspx (accessed 29 July 2024).

Government of Vietnam. 2011. *National Strategy against Climate Change.*

Ministry of Natural Resources and Environment. 2008. National Target Program to Respond to Climate Change: Implementing the Government's Resolution No. 60/2007/NQ-CP.

———. 2019. Tờ trình về dự án sửa đổi, bổ xung một số điều của Luật Bảo về Môi Trường.

National Assembly of the Socialist Republic of Vietnam. 2014. Luật bảo vệ môi trường. 55/2014/QH13.

Politbureau of the Communist Party of Vietnam. 1998. Chỉ thị về tăng cường công tác bảo vệ môi trường trong thời kỳ công nghiệp hóa, hiện đại hóa đất nước. 36/1998/CT-TW.

The President of the Socialist Republic of Vietnam. 2005. Law on Environmental Protection. 29/2005/L-CTN.

The Prime Minister of the Socialist Republic of Vietnam. 2004. Promulgating the Oriented Strategy for Sustainable Development in Vietnam (Vietnam's Agenda 21). 153/2004/GD-TTg.

Project VIE/89/021. 1989. Vietnam National Plan for Environment and Sustainable Development 1991–2000.

Office of the Prime Minister of the Socialist Republic of Vietnam. 2011. Quyết định: phê duyệt chiến lược quốc gia về biến đổi khí hậu. 2139/QĐ-TTg.

———. 2012a. Quyết định: Phê duyệt Đề án quản lý phát thải khí gây hiệu ứng nhà kính; quản lý các hoạt động kính doanh tín chỉ các-bon ra thị trường thế giới. 1775/QĐ-TTg.

———. 2012b. Quyết định: phê duyệt chiến lược quốc gia về tăng trưởng xanh. 1393/QĐ-TTg.

———. 2012c. Decision: Approval of the Green Growth Strategy. No. 1393/QĐ-TTg.

———. 2013. Quyết định: Phê duyệt Chiến lược quốc gia về đa dạng sinh học đến năm 2020, tầm nhìn đến năm 2030. 1250/QĐ-TTg.

Socialist Republic of Vietnam. 1994. Nghị định: của chính phủ số 175-CP ngày 18-10-1994 về hướng dẫn thi hành luật bảo vệ môi trường. 175-CP.

———. 1996. Nghị định của chính phủ số 26/CP ngày 26 tháng 4 năm 1996 quy định xử phạt vi phạm hành chính về bảo vệ môi trường.

———. 2012. "Implementation of Sustainable Development: Report at the United Nations Conference on Sustainable Development (RIO+20)".

———. 2015. Intended Nationally Determined Contribution of Viet Nam.

Working Group for the Reform of the Law for Environmental Protection. 2020. Suggestions for the Reform of the Law for Environmental Protection. Unpublished print version handed to the author in person.

NGO Sources

ActionAid Vietnam. n.d. Công việc chăm sóc không lương tái phân bổ để phát triển bền vững.

———. 2021. "Avoiding the Climate Poverty Spiral: Social Protection to Address Climate-Induced Loss & Damage".

CARE. 2019. Báo cáo thường niên.

CCWG. n.d. *Pathways for Policy Coherence in Implementation of NDC and SDGs in Viet Nam and the Role of Civil Society.*

———. n.d. Năng lượng Tái tạo và Hiệu suất Năng lượng tại Việt Nam: Hướng tới Sự bền vững.

———. n.d. "CCWG Position Paper on Co-benefits".

———. n.d. Các lộ trình liên kết chính sách trong triển khai đóng góp của quốc gia tự quyết định và các mục tiêu phát triển bền vững tại việt nam và vai trò của xã hội dân sự.

———. n.d. Câu chuyện thành công.

———. n.d. Thực hiên lời hứa: Thúc đẩy thích ứng với biến đổi khí hậu cho mọi lĩnh vực thành hiện thực tại Việt Nam.

———. n.d. "Climate Vulnerability Assessments by Civil Society Organizations: Towards the National Adaptation Plan (NAP) in Vietnam".

C&E. 2014. Hướng dẫn: giao đất giao rừng cộng đồng có sự tham gia.

———. 2015a. Tôi tham gia. Chuyện về quyền và trách nhiệm của cộng đồng dân tộc thiểu số đối với rừng tự nhiên miền trung Việt Nam.

———. 2015b. Báo cáo nghiên cứu: Lối sống sinh thái của sinh viên Việt Nam.

———. 2017. Tài liệu hỗ trợ lồng ghép lối sống sinh thái vào chương trình giáo dục cho thanh niên: Chủ đề tiêu dung thực phẩm bền vững.

———. 2018a. Tài liệu hỗ trợ lồng ghép lối sống sinh thái vào chương trình giáo dục cho thanh niên: Chủ đề làm vườn – dễ hay khó.

———. 2018b. Tài liệu hỗ trợ lồng ghép lối sống sinh thái vào chương trình giáo dục cho thanh niên: Chủ đề mua sắm.

CECR. 2015. "Assessing Women's Engagement in Environmental Impact Assessments on Infrastructure Projects in Vietnam: Recommendations for Policy and Public Participation in EIA".

———. 2017. Sổ tay hướng dẫn: sự tham gia của phụ nữ trong đánh giá tác động môi trường.

———. 2018. Báo cáo nghiên cứu: ô nhiễm nước và sự cần thiết phải xây dựng luật kiểm soát ô nhiễm nước tại Việt nam.

CENDI. 2014–15. "Livelihood Sovereignty and Village Wellbeing: H'rê People and the Spiritual Ecology".

———. 2019. "Bi-annual Narrative Report".

ENV. n.d. Hành động cấp bách: ngăn chặn sự tuyệt chủng của các loại động vật hoang dã. GreenID. 2016. Chiến lược phát triển 2016–2020.

Greenhub. n.d. "Monitoring and Assessment Programme on Plastic Litter in the Coastal Areas of Vietnam".

———. n.d. "Debris Data Collection and Monitoring System Design and Implementation".

———. 2020. "GreenBays Project: Moving toward Model Cities for Waste Management in Coastal Northeast Vietnam".

GreenID. 2016. *Chiến lược phát triển 2016-2020*.

———. 2018a. "Renewable Energy for All".

———. 2018b. "Policy Brief: Just Energy Transition – Opportunities and Challenges for Vietnam".

———. 2018c. Annual Report.

———. 2019. "Policy Brief Just Energy Transition: Opportunities and Challenges for Vietnam". Hanoi: Friedrich-Ebert-Stiftung.

Green Trees. n.d. 200 câu hỏi/đáp về môi trường.

———. 2016. "Báo cáo toàn cảnh tham hóa ước gửi tôi". GreenTrees website. https://www.greentreesvn.org/2016/10/bao-cao-toan-canh-tham-hoa-uoc-gui-toi.html (accessed 3 March 2021).

HEPA. 2020. Tọa đàm: nông nghiệp Sinh thái của Mạng lưới Nông dân nòng cốt giữa các tỉnh Cao Bằng, Hà Tĩnh, Quảng Bình, Kon Tum.

IPAM and Rosa-Luxemburg-Stiftung. 2019. Kỷ yếu tọa đàm quốc tế: Chia sẻ kinh nghiệm giữa Việt Nam và CHLB Đức về nghiên cứu và hoạch định chính sách trong bối cảnh chuyển đổi kinh tế, sinh thái và xã hội.

IUCN. 2018. Annual Review.

Live & Learn. 2011. Sổ Tay Truyền Thông Môi Trường.

Oxfam. n.d. "Vietnam Change Goal 3: Disaster Resilience". https://vietnam.oxfam.org/what-we-do/disaster-resilience (accessed 29 March 2021).

———. "Vietnam Change Goal 4. Resource Rights". https://vietnam.oxfam.org/what-we-do/resource-rights (accessed 29 March 2021).

———. 2019. "Voluntary Guidelines on Mitigating Socio-environmental Risks for Vietnamese Outward Investors in Agriculture in the Mekong Subregion".

———. 2020a. "Towards Sustainable Tax Policies in the ASEAN Region: The Case of Corporate Tax Incentives".

———. 2020b. Hướng tới chính sách thuế bền vững trong khối ASEAN: Trường hợp ưu đãi thuế thu nhập doanh nghiệp.

PanNature. n.d. "Publications". https://www.nature.org.vn/en/category/resources/publications/ (accessed 4 March 2021).

———. 2018. "Vietnam's Wildlife: Drained and Unsustained".

Schreurs, Miranda A., and Julia Balanowski. 2017. *Promoting Socially and Economically Just Energy Transformations in Asia: Possibilities, Challenges and the Road Ahead*. Hanoi: Friedrich-Ebert-Stiftung.

SNV. n.d. "REDD+ and Deforestation-Free Agriculture in Vietnam".

———. 2019. "Case Study: Mangroves and Market Projects (Phase I&II) in Vietnam".

SRD. 2018a. "Sustaining Our Momentum: Annual Report 2018".

———. 2018b. Tiếp tục vững bước: Báo cáo thường niên 2018.

VUSTA. 2018. Kinh tế xanh cho phát triển bền vững 2018: trong bối cảnh biến đổi khí hậu.

WWF. n.d. "Climate and Energy in Vietnam". WWF Vietnam. https://vietnam.panda.org/our_work_vn/climate_energy_vn/ (accessed 3 March 2021).

———. n.d. "A Better Future for Forests". Forest Solutions Platform. http://forestsolutions.panda.org/ (accessed 3 March 2021).

Index

350.org movement, 1

About the Author

Julia L. Behrens holds a PhD in Southeast Asian Studies from Humboldt University, Berlin. She has also studied at the University of Glasgow and spent time during her research at the University of Social Sciences and Humanities in Hanoi and at Yale University. Her research interests include social-ecological transformation, development, energy transitions, power structures, state-society relations and gender.

www.ingramcontent.com/pod-product-compliance
Lightning Source LLC
Chambersburg PA
CBHW042310210326
41598CB00041B/7341